Know them by their Fruit

First published by Jacana Media in 2021

10 Orange Street
Sunnyside
Auckland Park 2092
South Africa
+2711 628 3200
www.jacana.co.za

© A.T. Ankiewicz, 2021
Illustrations © Botanical Society of South Africa, 2021

All rights reserved.

ISBN 978-1-4314-3069-7

Illustrations by the author
Cover design by Aimèe Armstrong
Editing by Lara Jacob
Proofreading by Megan Mance
Set in Ehrhardt MT Std 11.5/15.5pt
Printed by Tandym
Job no. 003791

See a complete list of Jacana titles at www.jacana.co.za

Know them by their Fruit

A guide to identifying South African trees

A.T. Ankiewicz, Dip. For.

With a foreword by
Professor E.J. Moll

*Dedicated to the glory of the Creator,
to my wife and soulmate Jeanie,
and my parents who supported and encouraged me
for many years in completing this publication*

Contents

Help spread the plant message	vi
Foreword	ix
A brief history	xii
Author's note	xiv
How to use this book	xvi
Fruit key	xix
Colour plates	1
About the author	397
Acknowledgements	398
Index – botanical names	400
Index – common English names	405
Index – common Afrikaans names	409
Bibliography	413

Help spread the plant message

The world-renowned Kirstenbosch Botanical Garden in Cape Town was founded in 1913. A few months earlier, the Botanical Society of South Africa (BotSoc) was established to help support the Garden's aims, particularly spreading the message of the value of their collections and contribution to conservation. Today, BotSoc still aims to share the fascination of botanical gardens and inspire people from near and far to treasure and help protect South Africa's wonderfully diverse plant heritage.

Are you a member?

The Botanical Society of South Africa is committed to alerting everyone to help safeguard South Africa's indigenous heritage for the benefit of all who inhabit this earth. If you are not yet a member, we encourage you to join us and become part of this exciting endeavour. BotSoc has something for everyone who wants to discover more about our country's rich plant life that sustains our soul. To join, please visit https://botanicalsociety.org.za/.

In BotSoc, you will find a group of like-minded persons, walks and hikes that introduces you to the magic of nature, and how to join lectures and be active in conservation. Members often volunteer their time, skills and expertise to promote and protect our natural heritage for our own and future generations.

BotSoc members receive our quarterly magazine, *Veld & Flora*, which opens the gate to the secret garden called nature that is all around you wherever you may be. Depending on which membership category you choose, you can benefit from free admission to South African National Biodiversity Institute (SANBI) botanical gardens around South Africa.

Help spread the plant message

Support BotSoc publications

BotSoc also supports publications that spread and exchange knowledge about our country's extraordinary plants. We are able to help fund some publications and boost others through our 'BotSoc Recommended' logo.

Today, knowledge can be shared in several ways. Many of us enjoy the slower pleasures of the printed word, valuing the visual and tactile experiences it offers. Others prefer the versatility of digital platforms. BotSoc supports all these but appreciates that a cost always comes with generating text, illustrating, designing, printing and/or distributing.

We can help the authors, artists, printers, distributors and so on who make these glorious publications happen if you choose publications supported or recommended by BotSoc. Please also remember that you can help further through donations or bequests to the Botanical Society of South Africa, which are free of both donations and estate duty taxes.

Marinda Nel.

CHAIRPERSON, BOTANICAL SOCIETY OF SOUTH AFRICA

Foreword

As an ecologist and botanist, I see trees as pivotal to the way the general public perceives plants in the world around them.

The greater focus on South African trees started with the publication of *Sixty-Six Transvaal Trees* in 1966 – when the first list of national tree numbers was also published. This was when Trevor Ankiewicz's career in forestry and education started, so his journey parallels the expansion of our national tree list and many subsequent tree books.

Trees are, in a sense, flagship species that draw us into exploring both their rich diversity and, more broadly, the amazing biodiversity of plant life, particularly here in southern Africa. However, to novice plant lovers, that very diversity can seem so overwhelming because it becomes like a brick wall that stops their curiosity in its tracks. It is a barrier that we need to help them overcome so that we can foster the huge current interest in our indigenous flora and its identification among a growing number of citizen scientists. More and more people also grow trees and want to know more about them.

There are about 1 700 known indigenous species of trees in southern Africa, many of them mind-bendingly similar – somewhat akin to the birders' exasperation over birds that seem to be flocks of 'little brown jobs' (LBJs). The many vegetation regions in southern Africa mean that even the most experienced botanist will at some stage find themselves in a region where trees and other plants are unfamiliar. To the true plant-lover, that is part of the thrill of discovering for yourself the natural world – but having a well-informed guide gives you a head start.

That is how I first met Trevor, when he was our tour guide on an international dendrological Big 5 South Africa and Namibia trees tour in early 2018. On this tour he brought a preliminary mock-up of his book to show the international tree fundis. We in South Africa are indeed most fortunate to have so many excellent books on our flora, perhaps more and better than many other countries globally, but everyone was very impressed by Trevor's mock-up and commented that this was the first time they had seen a book where only fruit were used to identify trees.

We have many tree ID books that either cover the whole region, part of the region or focus on a particular genus. Most tree books use leaves as the main character for identification. They then add additional notes on other anatomical and morphological features to help the user confirm that identification. In Trevor's book, we now have a different companion guide to assist in this learning exercise. In my experience, this is the first and only book globally that focuses on the fruit of trees to enable one to identify species. For that reason alone, it is unique. I decided then and there that I would assist Trevor in ensuring the book's publication.

Through the Botanical Society of South Africa (BotSoc), and with the generous support of many people, this book was completed and Trevor Ankiewicz's dream of decades has come to fruition. In fact, he most generously donated the entire manuscript to BotSoc, waiving royalties and donating the 381 illustrations for BotSoc to sell, all of this because he was quite simply so grateful just to have the book made available to the interested public.

Trevor's journey started in his youth when he noticed that he was able to identify many trees confidently by their fruit. This started a life-long interest in tree-fruit identification and, of course, gave the focus of his book its different but intriguing character, the fruit. Now, after decades of carefully collecting, writing the informative and unusual text, and painstakingly illustrating fruiting twigs of 381 species, Trevor is at last able to share his work with all of us. He has also added some notes to assist with the identification, giving a personal sense by the author in this lovely book. I am sure that it will be prized by all who own and use it.

PROFESSOR E.J. MOLL, DEPARTMENT OF BIODIVERSITY & CONSERVATION BIOLOGY, UNIVERSITY OF THE WESTERN CAPE

Donors

Nicholson Trust

Sponsors

Marinda Nel
Lesley Cornish
Johan Barnard
Keith Kirsten
Nicholas Knott-Craig
Glen and Lucy Robbins

Subscribers

Botanical Society of Namibia
Eugene and Alice Moll
Kerin and Meyrick Bowker
Dave McDonald
Gary Hoile

A brief history

It was my father who first made me aware of trees when I was a young boy growing up on our highveld farm in the Bronkhorstspruit district of the former Transvaal. He taught me the names of American White Ash, Small-leaved Elm and Deodar Cedar. He showed me the difference between Black, Green and Silver Wattle and related how my grandfather had purchased their seed from the forestry department in Pretoria and had painstakingly planted them in situ and nurtured their seedlings in an otherwise treeless sea of grass. He taught me the value of their wood as a source of energy for the old kitchen stove that crackled and purred while it provided us with hot water, cozy warmth and making it possible for my mother to bake bread and prepare piping hot meals on cold winter days and nights. He made me feel the roughness of Black Ironbark and we would, on occasion, scratch through the pungent leaf litter and stringy bark to seek out the sweet white nuggets exuded by Manna Gum which we slowly savoured while leaning our backs against the towering smooth white trunks.

After completing my studies at the Saasveld School for Foresters just outside of George in the beautiful Garden Route in 1965, I requested a transfer to the South African Forestry Research Institute in Pretoria. Here, for a brief period, I was given the task of collecting the fruit of ornamental trees in the streets and cemeteries of the capital city for our seed store. This seed was destined for distribution to the various forestry nurseries and for purchase by the general public. It was during these collecting trips with my jovial team of men that it dawned on me, as a young inexperienced forester, that trees were more easily identified by their often striking fruit and pods than by their leaves and other features.

In 1969 I had the good fortune of being granted a transfer to the Horticultural Research Institute at Roodeplaat, north of Pretoria, which is situated in an almost pristine bushveld environment. Here I served as a research horticulturist and groundsman. After having worked at this wonderful institution for close to seven years, I made a brief, foolish and disastrous attempt at trying to manage a private nursery. The Department of Agricultural Technical Services regarded me worthy enough to be re-admitted into their service and I was duly appointed as a horticulturist at the world-renowned Botanical Research Institute in Brummeria, Pretoria (now SANBI). After a most stimulating seven years at this wonderful institution, I requested a transfer to the Forestry Research Centre at Saasveld near George to be closer to my then ailing mother-in-law. Here, as in my previous appointments, I served with some of the finest scientists and technicians in South Africa – and the world for that matter. After having worked here for four years, I was transferred to the regional office of the then Department of Nature and Environmental Conservation of the Cape Province Administration (a departmental name I still enjoy relating – now changed to CapeNature) where I completed my career as a civil servant amongst a dedicated pot-pourri of nature conservators and foresters who had a profound influence on my task as an environmental educator and public relations officer.

With the stimulation of the researchers, foresters and botanists at these various institutions, I decided in 1980 to embark on the mammoth task of compiling a book, which dealt solely with the fruit and pods of our indigenous trees. The idea behind this project was to supplement the existing information and publications on our trees and make them more recognisable to those who work with them, to those who study them and to those who just simply enjoy them.

With my father's influence, the environment in which I grew up in and an idea, which was conceived in the late 1960s, I was inspired to bring this guide to fruition. This then is my contribution to the conservation and creation of a greater awareness and appreciation of these magnificent plants that grace the many landscapes of this fair land.

Author's note

When I first started planning this book, my idea of a tree was a long-lived woody plant, with a sturdy developed trunk and an impressive crown of branches and leaves. These were initially the kinds of plants that I wished to concentrate on. Like some authors of tree guides, I did not regard most aloe species like the Mountain Aloe (*Aloe marlothii*) or the bushy Krantz Aloe (*Aloe arborescens*) – despite its specific name – as trees in the true sense of the word. Our beautiful cycads and tree ferns are, to my mind, also not included here as real trees. In my travels I have yet to come across the colourful Cape Honeysuckle (*Tecomaria capensis*) and that delightful Pride of De Kaap (*Bauhinia galpinii*), which grows in the Barberton valley, having developed as a shady, truly recognisable tree.

As the book developed, this distinction between what I regarded as a tree and what I regarded as a shrub became more and more blurred. I came to realise that habitat and climate greatly influence the stature and development of these plants. A classic example here is the ubiquitous Sweet Thorn (*Acacia karroo* – now *Vachellia karroo*), which occurs as a stunted bush in the dry river courses of the Great Karoo, yet develops into an impressive tree with a sturdy black bole and rounded crown in the Mpumalanga bushveld.

Leaves, I found, tend to be as variable as the growth habit of trees in different situations. Leaves tend to be larger in a shaded spot than where they are exposed to full sun. The sizes of leaves could also vary considerably on a tree species growing on a riverbank and on a wet site as compared to one situated on a dry rocky koppie. Juvenile foliage is, in some instances, quite different from mature leaves on the same tree.

These variables, I have found, sometimes even confuse the trained botanist, no matter how well the leaf key has been designed. Unlike the size of the leaves, or of the tree itself, my many observations throughout the country have revealed that the fruit and pods of our native trees, with a few exceptions, vary little in size, shape and colour from one place to another. That is where this book should be invaluable in making it relatively easy for the lay person to confidently identify many of our trees wherever they might be growing.

I have attempted to design a very user-friendly fruit key, which will guide the reader to specific groups of fruit and pods depicted in the colour plates. Once the reader has found the right group of illustrations in the key, a quick page through the relative colour plates section should reveal the fruit (and tree) in question, provided, of course, that it is featured in this book in the first place.

This may seem rather basic but I have heard so many people identifying seeds as fruit. To avoid any confusion, readers must please note that there is a distinct difference between a fruit and a seed: an apple is regarded as a fruit, but the pips within are, in actual fact, the seeds, and it is important to bear this in mind. Pods are also fruit but they have been dealt with separately in the fruit key.

Another feature of this book, which I have not yet come across in other literature on trees, is an illustration of the partially decomposed fruit or the exposed seeds after they have lain in the leaf litter below the tree for several weeks or even months. Fruiting time is often of short duration and confined to different times of the year for different species of trees. It can, however, be said that many fruit, pods or their seeds will persist beneath the tree for a great deal longer than what they did on the tree itself. By seeking out the dry remains of the fruit, which may persist on the tree or the husks or seeds on the ground, and by consulting this book, the tree enthusiast will often be able to positively identify a tree long after its fruiting time has passed and irrespective of what time of year it might be or whether the tree is in leaf or not.

Bear in mind that many of our tree species are dioecious (male and female reproductive organs are not found on one tree but occur on separate trees), which means that fruit or pods will only be found on female trees.

The fruit and pods of exotics (trees introduced from other parts of Africa or the world) may appear very similar to some of our native trees and be confusing to the reader; however, taking note of the distribution of the tree in question, leaf size, their arrangement on the branch and the nature of the venation should allay any further doubt in this regard.

I have deliberately included long-standing botanical names (as synonyms, and they appear in square brackets) in this publication as most of the literature (post 1977) dealing with trees has used these names for a very long period of time and I believe that this decision will contribute to the present usefulness of this book and will promote the conservation of our trees rather than confuse the reader with all the taxonomic revision that has taken place in recent years.

Common names may differ from region to region as well as in the many books that have already been published on our native trees.

Additional note from Professor E.J. Moll, retired professor of botany and tree hugger:

There is a wonderfully short and most appropriate line in Bill Bryson's A Short History of Nearly Everything about taxonomists, in which Bryson describes them as being both scientists and artists – waging war with one another.

In chapter 23, 'The richness of being', Bryson writes eloquently, and with such well-directed humour, about the vagaries of taxonomy and the people who practise the science/art.

His prose is so poignant for us here in South Africa, where we, along with 127 other countries, lost the privilege to use the name 'acacia' for our iconic thorn trees in a long and protracted battle where one Australian and one British taxonomist used their sly skills to rob us of the ancient name.

How to use this book

This book is divided into two sections, namely fruit keys and colour plates:

1. Fruit keys: These keys have been designed to show the shapes, sizes and other diagnostic features of fruit and pods. The key will guide the reader to a specific group in the second section of the book. These are arranged first by size of round/roundish fruits (see below), then by other fruit shapes that are neither round nor oval:

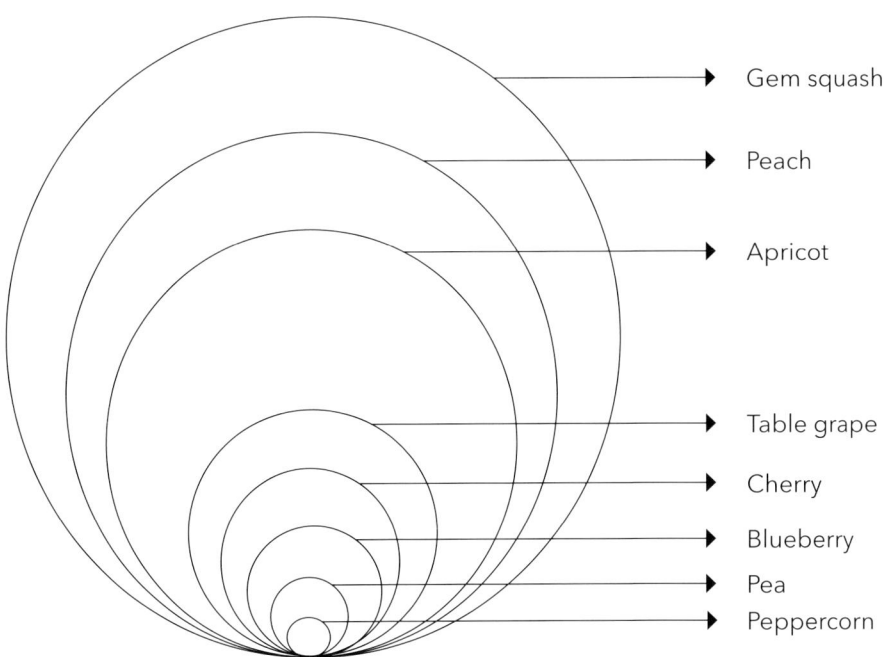

Know them by their Fruit

2. Colour plates: The colour plates, with the exception of a few fruit and pods that were too large to be accommodated in the book's format, were illustrated as closely as possible to actual size. In instances where these fruits and pods have been reduced, it will be noted as such below each illustration.

Some illustrations have been duplicated due to fruit size differences in different collection localities.

Both botanical and common names (English and Afrikaans) are given.

Below each illustration is pertinent information about the tree and its fruit. The table below is indicative of the information and lists the various elements of that information as they appear.

Line 1	SA tree number + scientific name + synonym + English common name + Afrikaans common name	392 | *Searsia pyroides* [=*Rhus pyroides*] | Common wild currant | Gewone taaibos
Line 2	Distribution + fruiting time	Widespread except in the dry west | November to May
Line 3	Habitat	Open woodland, dry thornveld and frequently associated with Vachellia karroo
Line 4	Description	A small tree, twigs with velvety hairs; leaves thin, dull olive-green

Fruit keys

Peppercorn size fruit
(Plates 1 to 24)

Peppercorn size

- **soft, fleshy, leathery**
  - **smooth shiny**
    - clustered or bunched → round pl. 1-9, oval pl. 21-22
    - not clustered or bunched → round pl. 10
  - **rough hairy velvety**
    - clustered or bunched → round pl. 11-12
  - **flattened**
    - not clustered or bunched
      - not splitting → round pl. 13
      - splitting 2x → round pl. 14–15
    - clustered
      - splitting 2x → round pl. 17
- **hard woody**
  - **smooth shiny**
    - not clustered or bunched
      - not splitting → round pl.16
      - splitting 3x → round pl. 18-19
  - **rough hairy velvety**
    - clustered or bunched
      - splitting 4x or more → oval pl. 23
      - (oval pl. 24)
  - **bumpy warty**
    - not clustered or bunched
      - splitting 2x → round pl. 20

Pea size fruit
(Plates 25 to 72)

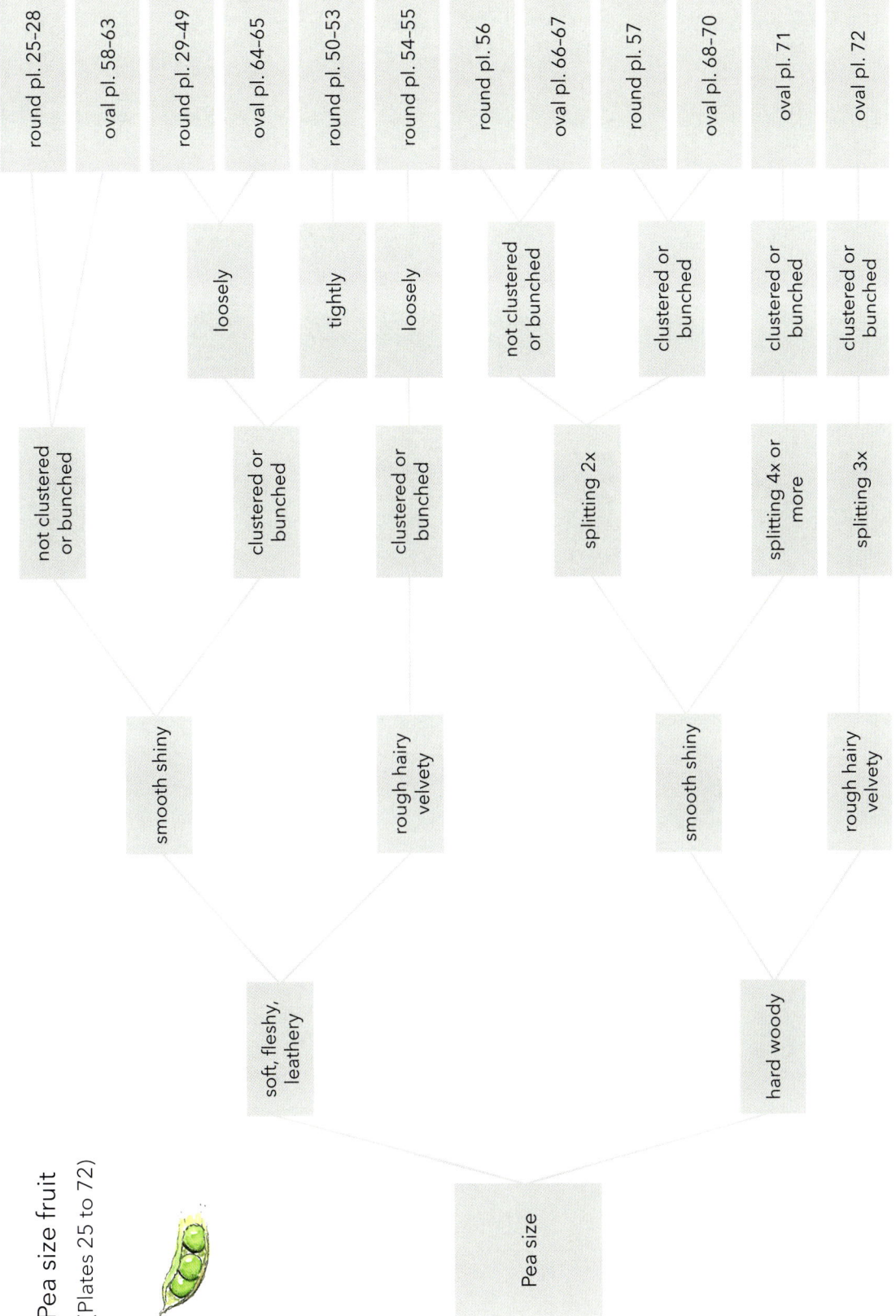

Blueberry size fruit
(Plates 73 to 126)

Blueberry size

- **soft, fleshy, leathery**
  - **smooth shiny**
    - not clustered or bunched → round pl. 73–80
    - not clustered or bunched → oval pl. 102–110
    - clustered or bunched → round pl. 81–95
    - clustered or bunched → oval pl. 111–120
  - **rough hairy velvety**
    - not clustered or bunched → round pl. 96–98
    - clustered or bunched → oval pl. 121
  - **bumpy warty**
    - clustered or bunched → round pl. 99
- **hard woody**
  - **smooth shiny**
    - not splitting → oval pl. 122
    - splitting 3x
      - clustered → oval pl. 123
      - not clustered or bunched → oval pl. 124–125
    - splitting 4x or more
      - clustered or bunched → round pl. 100–101
      - not clustered or bunched → oval pl. 126

Cherry size fruit
(Plates 127 to 172)

```
Cherry size
├── soft, fleshy, leathery
│   ├── smooth shiny
│   │   ├── not clustered or bunched
│   │   │   ├── round pl. 127-134
│   │   │   └── oval pl. 154-160
│   │   └── clustered or bunched
│   │       ├── round pl. 135-145
│   │       └── oval pl. 161-167
│   └── rough hairy velvety
│       ├── not clustered or bunched
│       │   ├── oval pl. 168-169
│       └── clustered or bunched
│           ├── round pl. 146-149
│           └── oval pl. 170
└── hard woody
    ├── smooth shiny
    │   ├── splitting 2x
    │   │   ├── not clustered or bunched → round pl. 150
    │   └── splitting 3x
    │       ├── not clustered or bunched → oval pl. 171
    │       └── clustered or bunched → round pl. 151-152
    └── rough hairy velvety
        ├── splitting 3x
        │   └── clustered or bunched → round pl. 172
        └── splitting 4x or more
            └── clustered → round pl. 153
```

Table grape size fruit
(Plates 173 to 206)

Table grape size

- **soft, fleshy, leathery**
  - **smooth shiny**
    - **not clustered**
      - round pl. 173–179
      - oval pl. 194–199
    - **clustered**
      - round pl. 180–182
      - oval pl. 200–202
  - **rough hairy velvety**
    - **not clustered**
      - round pl. 183
    - **clustered**
      - round pl. 184–187
- **hard woody**
  - **smooth shiny**
    - **not splitting**
      - round pl. 188
      - oval pl. 203
    - **splitting 2x**
      - not clustered — oval pl. 204–205
    - **splitting 3x**
      - not clustered — round pl. 189
    - **splitting 4x or more**
      - not clustered — round pl. 190–191
  - **rough hairy velvety**
    - **not splitting**
      - not clustered — oval pl. 206
    - **splitting 4x or more**
      - not clustered — round pl. 192–193

Apricot size fruit
(Plates 207 to 227)

```
Apricot size
├── soft, fleshy, leathery
│   ├── smooth shiny
│   │   ├── not clustered ── round pl. 207–210
│   │   └── clustered ── oval pl. 213–214
│   ├── rough hairy velvety
│   │   ├── clustered ── round pl. 211–212
│   │   └── not clustered ── oval pl. 215–216
│   └── bumpy warty
│       ├── oval pl. 218
│       └── oval pl. 219
└── hard woody
    ├── smooth shiny
    │   ├── not splitting
    │   │   ├── round pl. 220
    │   │   └── oval pl. 226
    │   ├── clustered
    │   │   ├── round pl. 221
    │   │   └── oval pl. 227
    │   └── splitting 4x or more ── round pl. 222–223
    └── rough hairy velvety
        ├── not splitting ── round pl. 217
        └── splitting 3x ── round pl. 224–225
```

Peach size fruit
(Plates 228 to 230)

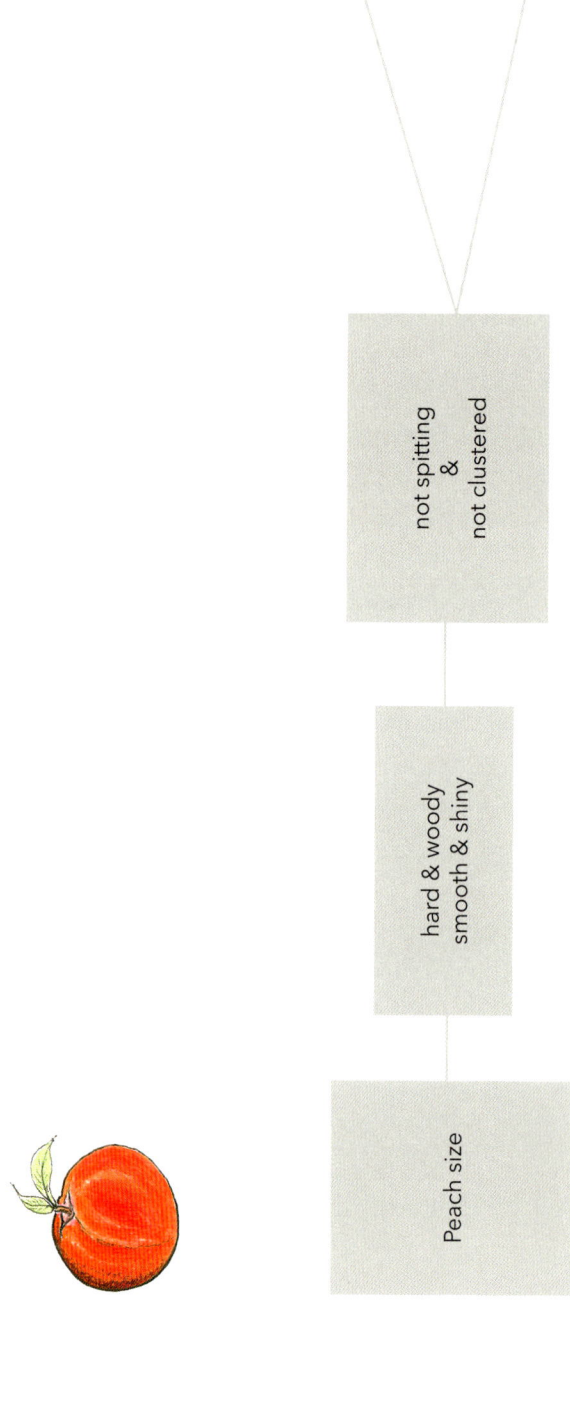

Peach size → hard & woody smooth & shiny → not spitting & not clustered →
- round pl. 228–229
- oval pl. 230

Gem squash size fruit
(Plates 231 to 235)

- Gem squash size
  - soft, fleshy, leathery
    - smooth shiny
      - not clustered
        - round pl. 231
  - hard woody
    - smooth shiny & not splitting
      - not clustered
        - round pl. 232-233
    - rough hairy velvety & not splitting
      - not clustered
        - round pl. 234
        - oval pl. 235

Pods (straight)

(Plates 236 to 267)

- **Pods**
  - **short straight**
    - **slender**
      - **papery leathery**
        - not splitting pl. 236-239
        - splitting pl. 240
      - **cardboard like**
        - not splitting pl. 241
        - splitting pl. 242-243
    - **broad**
      - **papery leathery**
        - not splitting pl. 244
        - splitting pl. 245-246
      - **cardboard like**
        - not splitting pl. 247-248
        - splitting pl. 249
      - **woody**
        - not splitting pl. 250-253
        - splitting pl. 254-255
  - **long straight**
    - **slender**
      - **papery leathery**
        - splitting pl. 256-258
      - **cardboard like**
        - splitting pl. 259
      - **woody**
        - splitting pl. 260
    - **broad**
      - **papery leathery**
        - not splitting pl. 261
        - splitting pl. 262-263
      - **cardboard like**
        - splitting pl. 264-265
      - **woody**
        - not splitting pl. 266
        - splitting pl. 267

Pods (other)
(Plates 268 to 285)

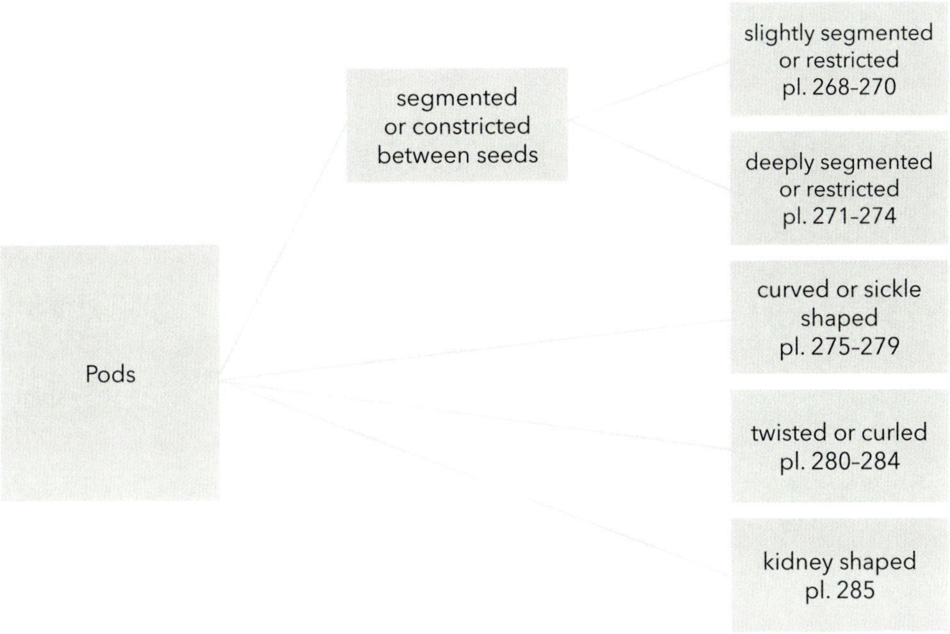

- Pods
  - segmented or constricted between seeds
    - slightly segmented or restricted
      pl. 268-270
    - deeply segmented or restricted
      pl. 271-274
  - curved or sickle shaped
    pl. 275-279
  - twisted or curled
    pl. 280-284
  - kidney shaped
    pl. 285

Winged/lobed/cylindrical
(Plates 286 to 334)

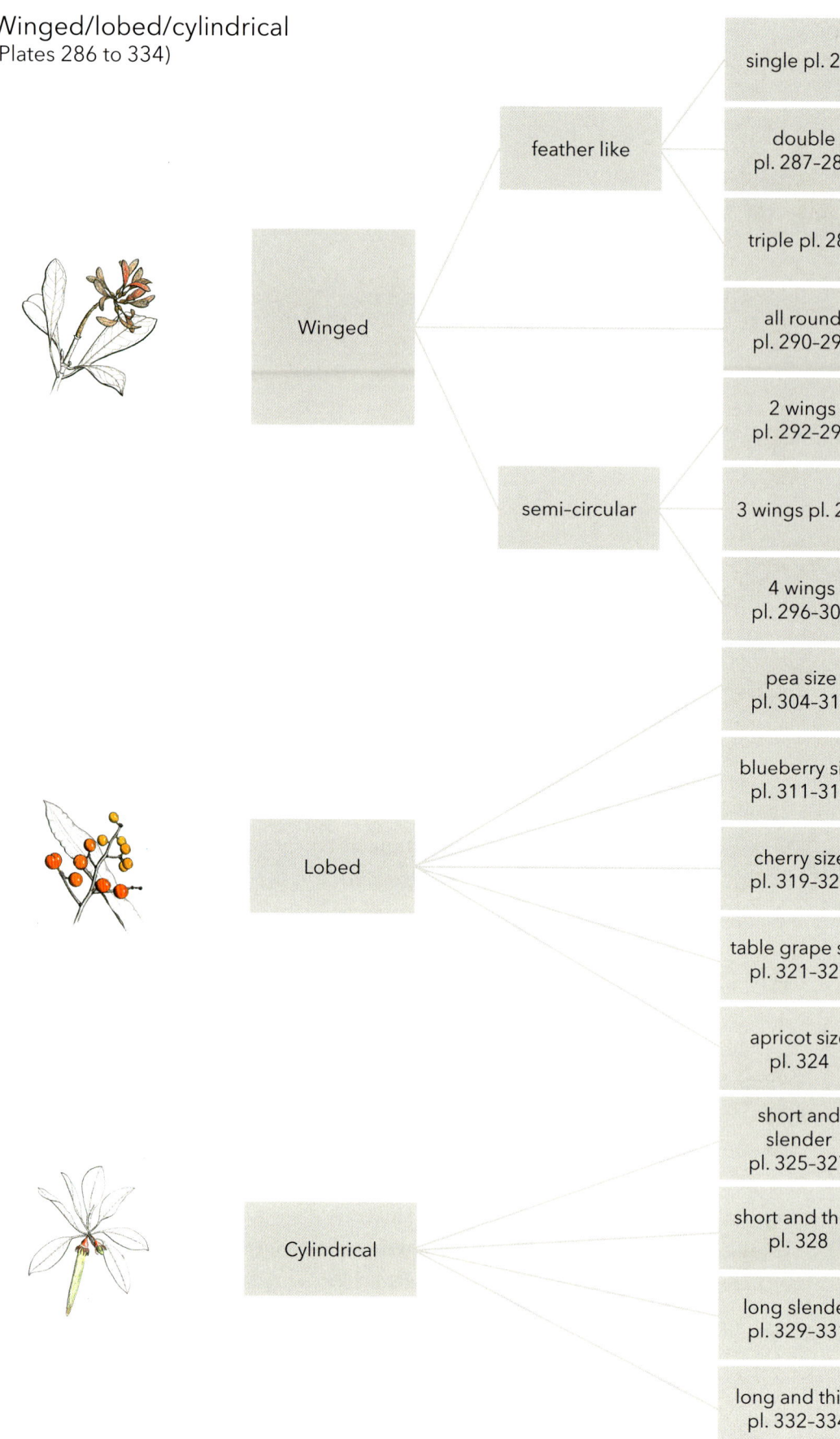

- **Winged**
  - **feather like**
    - single pl. 286
    - double pl. 287–288
    - triple pl. 289
  - all round pl. 290–291
  - **semi-circular**
    - 2 wings pl. 292–294
    - 3 wings pl. 295
    - 4 wings pl. 296–303
- **Lobed**
  - pea size pl. 304–310
  - blueberry size pl. 311–318
  - cherry size pl. 319–320
  - table grape size pl. 321–323
  - apricot size pl. 324
- **Cylindrical**
  - short and slender pl. 325–327
  - short and thick pl. 328
  - long slender pl. 329–331
  - long and thick pl. 332–334

Other
(Plates 335 to 381)

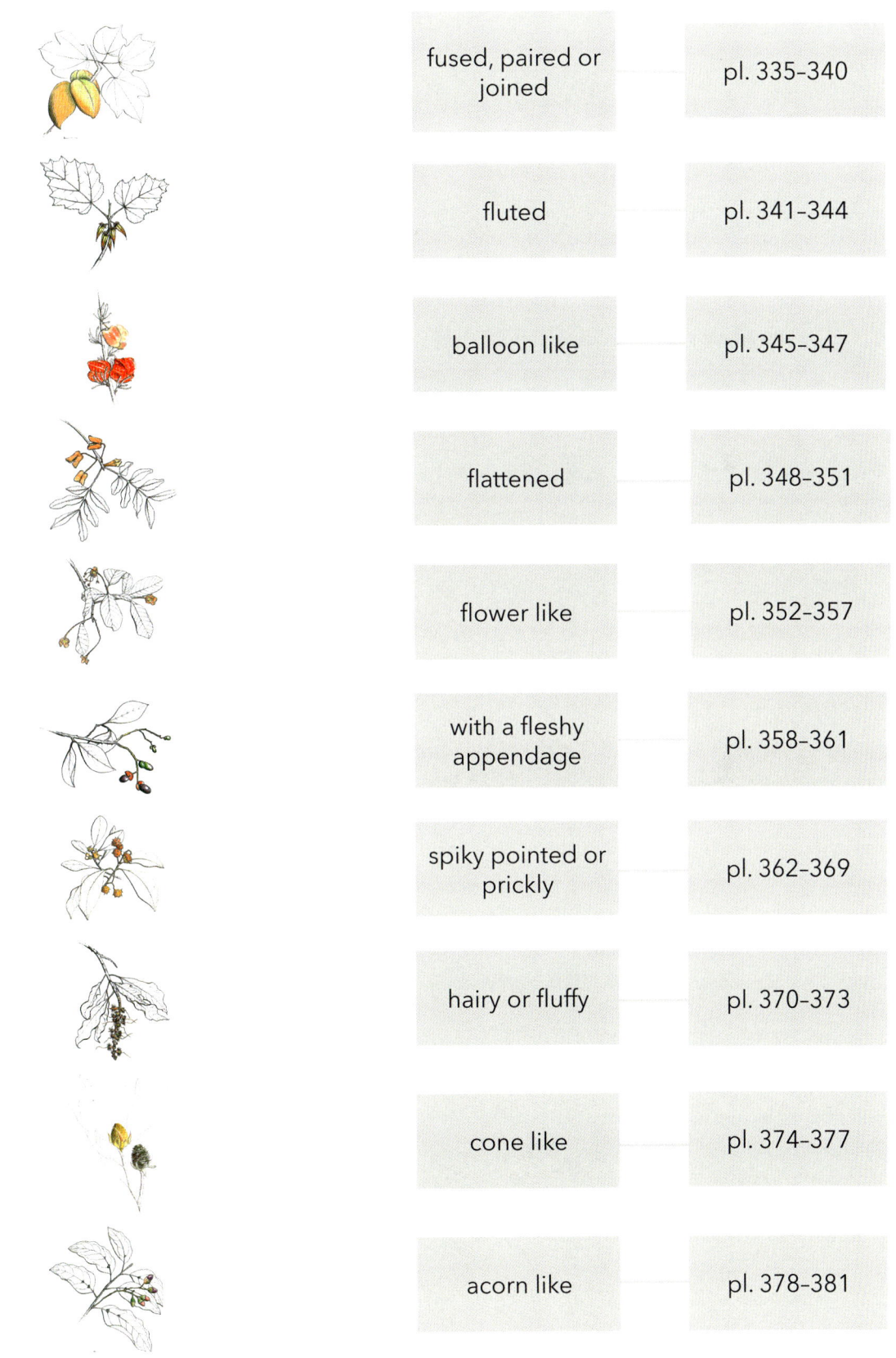

fused, paired or joined	pl. 335–340
fluted	pl. 341–344
balloon like	pl. 345–347
flattened	pl. 348–351
flower like	pl. 352–357
with a fleshy appendage	pl. 358–361
spiky pointed or prickly	pl. 362–369
hairy or fluffy	pl. 370–373
cone like	pl. 374–377
acorn like	pl. 378–381

Peppercorn size fruit

(Plates 1 to 24)

PLATE 1

392 | *Searsia pyroides* [=*Rhus pyroides*] | Common wild currant | Gewone taaibos
Widespread except in the dry west | November to May
Open woodland, dry thornveld and frequently associated with *Vachellia karroo*

A small tree, twigs with velvety hairs; leaves thin, dull, olive-green

PLATE 2

388.1 | *Searsia lucida* [=*Rhus lucida*] | Glossy currant | Blinktaaibos
Mpumalanga, north-east Free State, KwaZulu-Natal, coastal Eastern and Western Cape | October to November
Bush clump edges, scrub forest and protected places in grassland

Leaves are 'sticky', shiny and hairless; fairly leathery and dark olive-green

PLATE 3

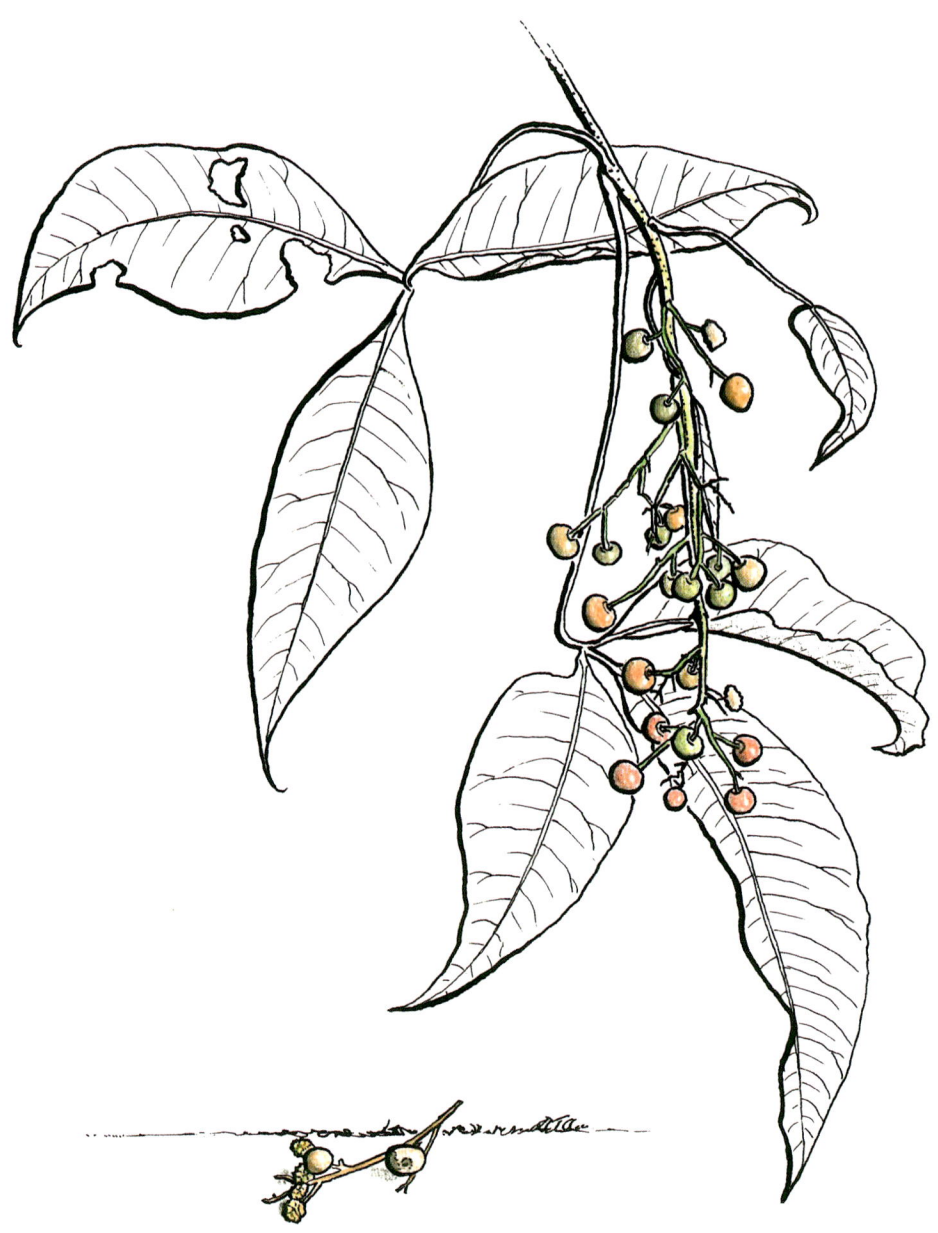

380 | *Searsia chirindensis* [=*Rhus chirindensis*] | Red currant | Bostaaibos
Mpumalanga, eastern KwaZulu-Natal, Eastern Cape to the Overberg |
November to February
Rocky hillsides, open woodland, riverine thicket, inland forest

Biggest of the Searsia *spp.; leaves pendent, petioles turning red before falling*

PLATE 4

397 | *Ilex mitis* | Cape holly | Without
Widespread in central and eastern South Africa to Cape Town | April to July
Evergreen forest, river banks

Petiole and main vein channeled above; leaf margins toothed towards the apex

PLATE 5

699 | *Empogona lanceolata* [=*Tricalysia lanceolata*] | Jackal coffee | Jakkalskoffie
Gauteng, Mpumalanga, eastern KwaZulu-Natal, north-east Eastern Cape | December to March
Forest, forest margins, river fringes, wooded ravines

Leaves hairless, margin wavy, venation more-or-less translucent below

PLATE 6

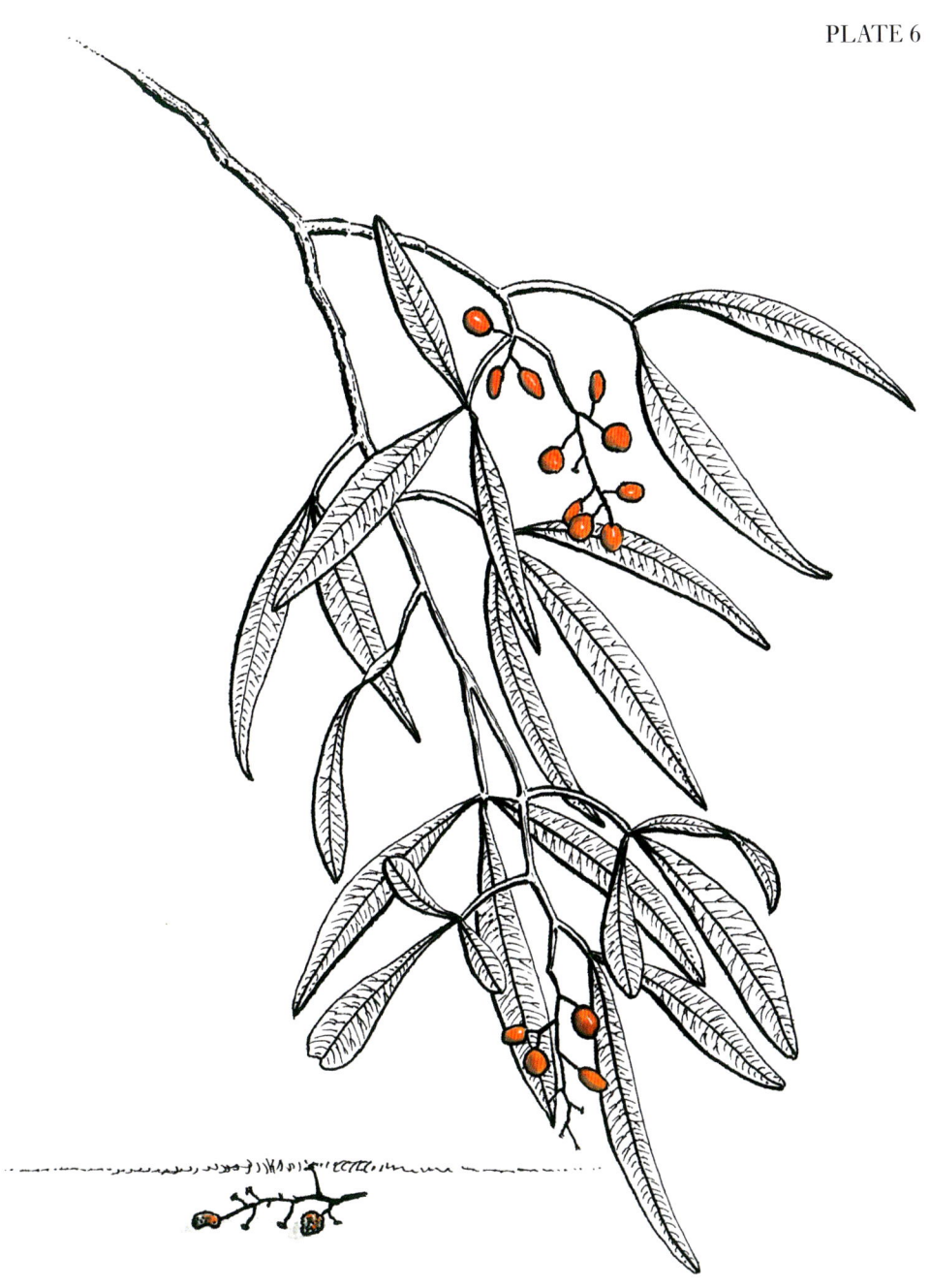

396 | *Searsia pendulina* [=*Rhus pendulina*] | White karee | Witkaree
Northern Cape | December to May
Orange (Gariep) river banks and in adjacent arid semi-desert

Young stems can be spine like; leaves tend to hang and are a dull green.

PLATE 7

42 | *Trema orientalis* | Pigeonwood | Hophout
Soutpansberg, Magaliesberg, eastern KwaZulu-Natal and Eastern Cape to the Garden Route | January to June
Forest margins, riverine bush

Leaf margins finely toothed; **Celtis africana**-*like, leaf margins only on the bottom half*

PLATE 8

658 | *Premna mooiensis* | Skunk-bush | Muishondbos
Eastern Mpumalanga, eastern Limpopo, Kruger National Park, Eastern Cape | November to December
Rocky outcrops, thicket

Usually a small tree; crushed leaves with a strong smell; bark flakes in strips

PLATE 9

566 | *Schefflera umbellifera* | False cabbage tree | Basterkiepersol
Eastern Limpopo, Mpumalanga | April to August
Coastal and mountain forest in KwaZulu-Natal and Eastern Cape

Large, pinnately compound leaves are a give-away.

PLATE 10

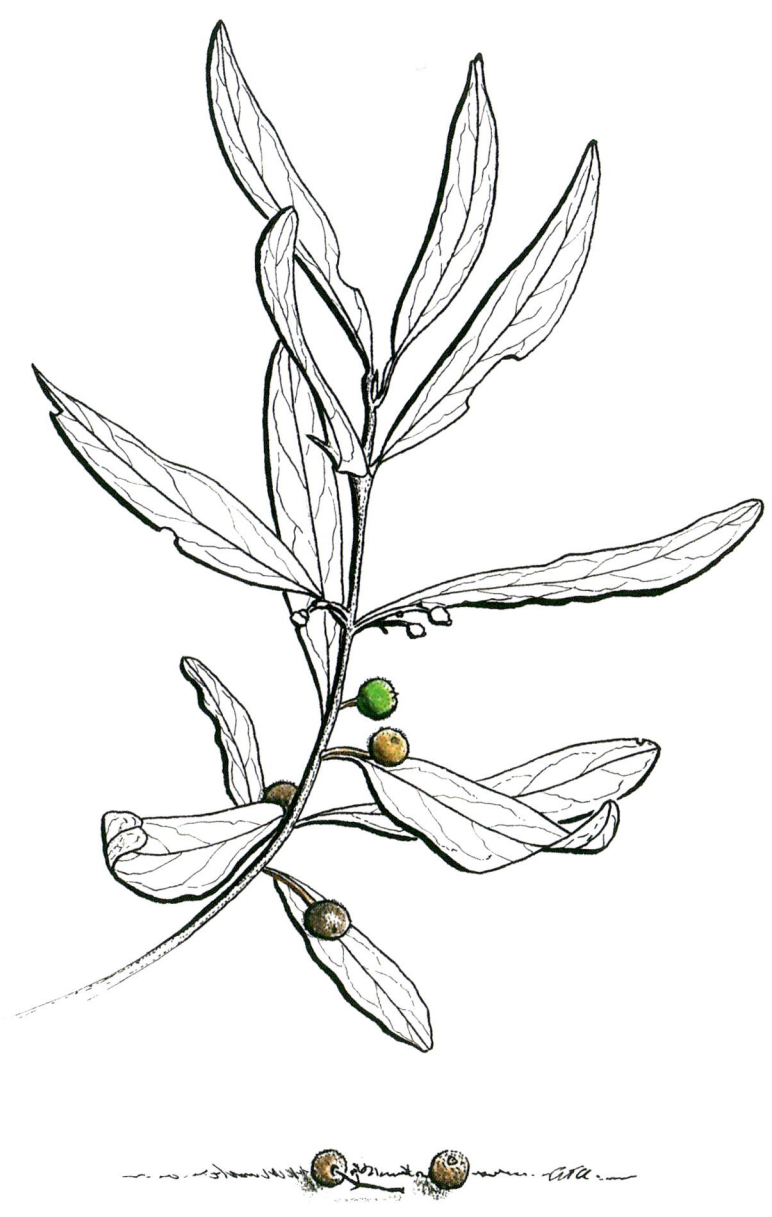

594 | *Euclea crispa* | Blue guarri | Bloughwarrie
Widespread except in drier west | April to December
Open woodland, forest margins, rocky outcrops

Tough, stiff, leathery; leaves with a sometimes wavy margin are characteristic.

PLATE 11

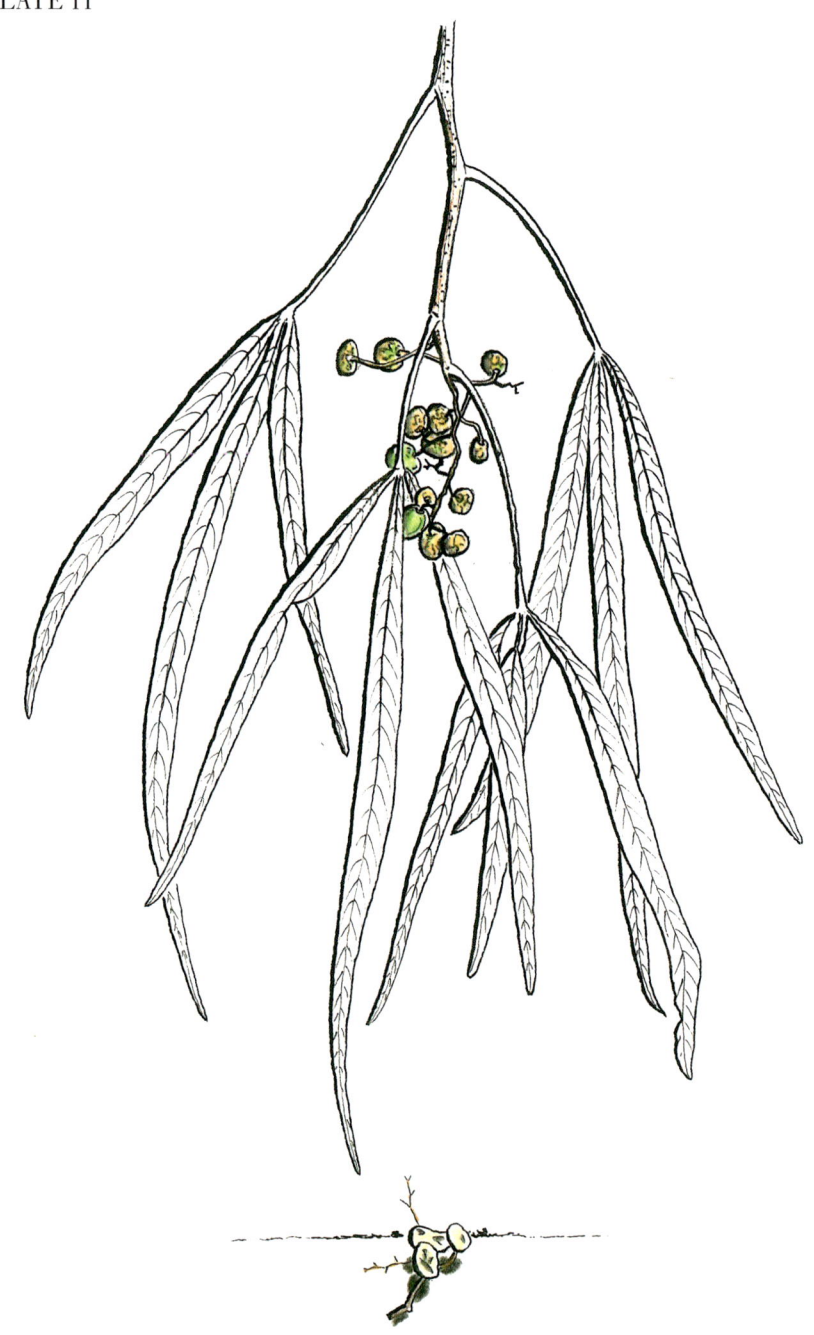

386 | *Searsia lancea* [=*Rhus lancea*] | Karee | Karee
Free State, central Eastern and Western Cape, North West, northern Northern Cape | September to January
Open woodland, river and stream banks

Pendent leaves and branches (weeping willow-like) and rough brown bark

PLATE 12

387 | *Searsia leptodictya* [=*Rhus leptodictya*] | Mountain karee | Bergkaree
North West, Limpopo, Gauteng, Mpumalanga, Free State | March to May
Open woodland, grassland, forest margins

Similar to the Karee but leaves not as long and margins serrated

PLATE 13

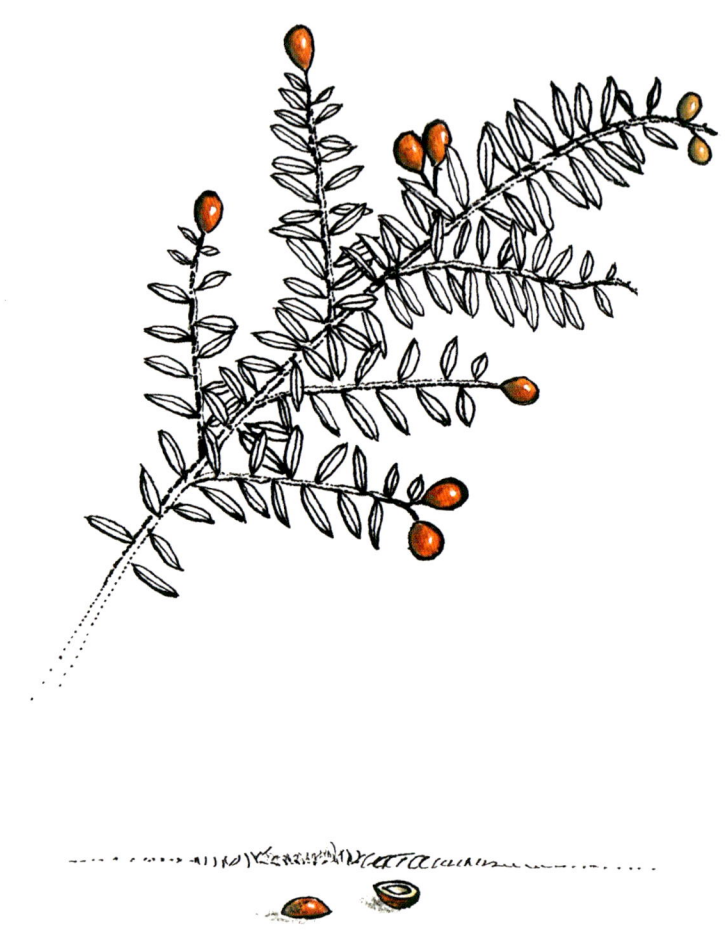

453.2 | *Phylica paniculata* | Common hard-leaf | Gewone hardeblaar
Western Limpopo, Mpumalanga, Soutpansberg in a wide arc to the Overberg | August to September
Rocky mountain slopes, riverine bush except on the coast

The tough leaves are grey-green above and white below.

PLATE 14

253 | *Zanthoxylum capense* [=*Fagara capensis*] | Small knobwood | Kleinperdepram
North West, Gauteng, Limpopo, Mpumalanga, KwaZulu-Natal, Eastern Cape, Garden Route | March to April
Dry woodland, forest margins, rocky outcrops

Straight prickles on knobs of older wood; crushed leaves have a citrus-like smell.

PLATE 15

335 | *Macaranga capensis* | Wild poplar | Wildepopulier
Coastal KwaZulu-Natal, north-eastern Eastern Cape | March to April
Evergreen forest, forest fringes, stream banks, marshland

Petiole attached to large leaves from below; margins entire or serrated

PLATE 16

523 | *Galpinia transvalica* | Wild Pride of India | Bosveld Liguster
Eastern Limpopo, north-eastern Mpumalanga | June to July
Woodland and rocky outcrops

The mid-vein ending in a knob-like gland is diagnostic.

PLATE 17

144 | *Trichocladus grandiflorus* | Green witch-hazel | Groen towerhaselaar
North-eastern Limpopo, Mpumalanga, north-eastern KwaZulu-Natal |
February to May
Mist belt forest fringes, wooded ravines

Leaves paler below with veins prominent; new leaves reddish and hairy

PLATE 18

399 | *Gymnosporia buxifolia* [=*Maytenus heterophylla*] | Common spike-thorn | Gewone pendoring
Widespread | May to January
Wide range, forest margins, a pioneer

Leaves variable in shape with strong fearsome spine-tipped branches

PLATE 19

455 | *Heteropyxis natalensis* | Lavender tree | Laventelboom
Eastern North West, Limpopo, eastern KwaZulu-Natal | March to May
Bushveld, forest margins

Flaking bark with new patches yellowish; leaves aromatic when crushed

PLATE 20

336.1 | *Clutia pulchella* | Warty-fruited lightning-bush | Gewone bliksembos
Widespread from Limpopo down eastern half of South Africa to Cape Town | January to May
Karoo scrub to evergreen forest

Usually a small tree (even a shrub) with the occasional orange-reddish leaf

PLATE 21

498 | *Scolopia zeyheri* | Thorn pear | Doringpeer
Widespread in eastern half of South Africa down to Swellendam | July to September
Evergreen forest and forest margins, often associated with termite mounds

Can have massive spines on the trunk; crackled leaves reveal a milky waxy layer

PLATE 22

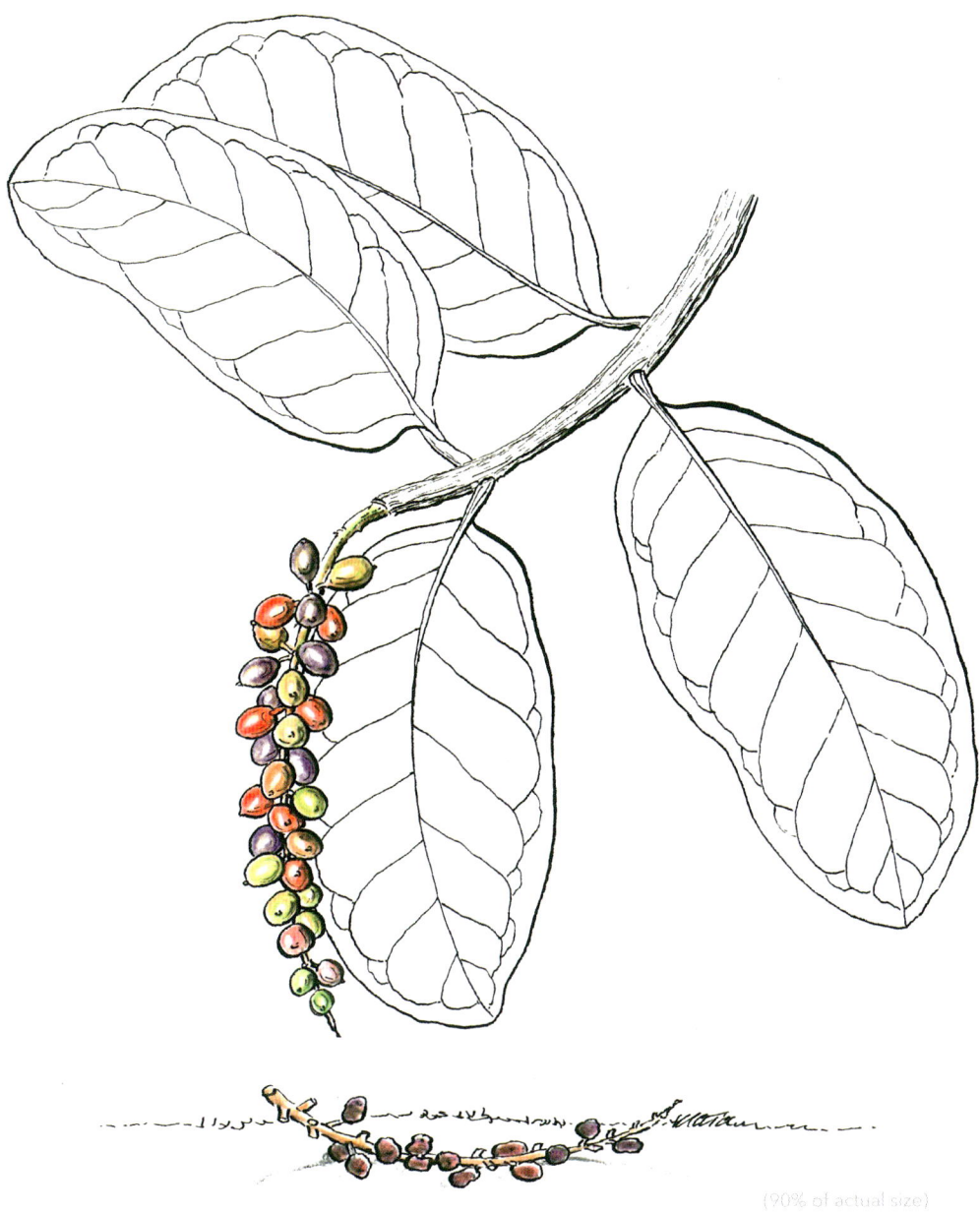

(90% of actual size)

318 | *Antidesma venosum* | Tasselberry | Tosselbessie
Eastern Limpopo, eastern Mpumalanga, coastal KwaZulu-Natal, coastal eastern Eastern Cape | March to May
Coastal bush, moist bushveld, forest margins

Leaves hairy below; veins looping; fruit can be almost pea size

PLATE 23

503 | *Trimeria grandifolia* [=*Trimeria rotundifolia*] | Wild mulberry | Wildemoerbei
Eastern Limpopo, Mpumalanga, KwaZulu-Natal, coastal Eastern Cape to the Garden Route | February to April
Forest, forest margins, wooded valleys

Big leaves with three or more veins from the base and serrated margins are characteristic.

PLATE 24

634 | *Nuxia floribunda* | Forest elder | Bosvlier
Northern Limpopo, Mpumalanga, coastal KwaZulu-Natal, Eastern Cape, Western Cape, Garden Route | June to October
Evergreen forest, forest margins

Leaves in threes; longish, pinkish petioles and drooping leaves are characteristic.

Pea size fruit
(Plates 25 to 72)

PLATE 25

500 | *Gerrardina foliosa* | Krantz berry | Kransbessie
Eastern Limpopo, Mpumalanga, eastern KwaZulu-Natal, north eastern Eastern Cape | March to September
Forest margins, rocky outcrops, mist belt, forested ravines

Small erect trees on sandstone areas; bright green leaves with serrated margins

PLATE 26

39 | *Celtis africana* | White stinkwood | Witstinkhout
Widespread except in dry arid areas | October to December
Coastal forest, bushveld, grassland, rocky outcrops

Leaves hairy; prominent basal veins, unequal leaf base and trees conspicuously deciduous-grey in winter

PLATE 27

452 | *Rhamnus prinoides* | Dogwood | Blinkblaar
Widespread from Limpopo down inland eastern South Africa to the Overberg | January to March
Forest margins, riverine forest

Shiny leaves; veins depressed above, prominent below; margins finely toothed

PLATE 28

601 | *Euclea undulata* | Common guarri | Gewone ghwarrie
Widespread except central Free State, KwaZulu-Natal and West Coast |
April to October
Open woodland, rocky ridges, grassland, semi-desert

Small neat trees with tough leaves that have a very undulating margin

PLATE 29

381 | *Searsia dentata* [=*Rhus dentata*] | Nana berry | Nanabessie
Widespread in eastern half of South Africa | November to January
Forest margins, thornveld, rocky hillsides, open woodland

The deeply incised (toothed) leaflets with veins raised above are characteristic.

PLATE 30

147 | *Prunus africana* | Red stinkwood | Rooistinkhout
Eastern Limpopo, eastern Mpumalanga, central KwaZulu-Natal, inland Eastern Cape | September to January
Evergreen and mist belt montane forest

Rough brown blocky bark is diagnostic; crushed leaves have a faint almond smell.

PLATE 31

423 | *Allophylus decipiens* | False currant | Bastertaaibos
Mpumalanga, central KwaZulu-Natal, Eastern Cape coast, Garden Route |
April to June
Coastal forest, riverine fringe forest, wooded ravines

Similar to **Searsia** *spp. [=*Rhus*] but has hairy pockets in axils of the veins below*

PLATE 32

718.1 | *Pavetta lanceolata* | Forest bride's bush | Bosbruidsbos
Mpumalanga, KwaZulu-Natal, coastal Eastern Cape | March to August
Coastal evergreen forest, inland forest, bushveld

All **Pavetta** *spp. have small black nodules in the leathery leaf blade.*

PLATE 33

496 | *Scolopia mundii* | Red pear | Rooipeer
Mpumalanga, KwaZulu-Natal, coastal East and Western Cape | October to January
Mountain evergreen forest, forest margins

Leaves stiff, shiny; margins with sharp serrations, black-tipped towards the base

PLATE 34

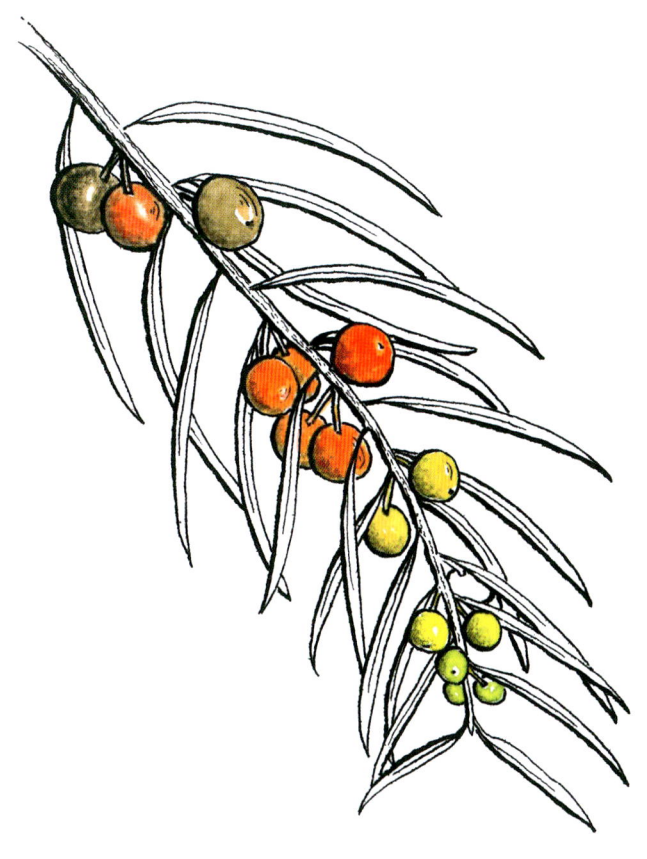

598 | *Euclea pseudebenus* | Ebony tree | Ebbeboom
Arid north-western Northern Cape | February to May
Desert, semi-desert, along water courses

This is the only Euclea *that looks like a weeping willow.*

37

PLATE 35

414 | *Cassine peragua* | Cape saffron | Bastersaffraan
KwaZulu-Natal, Eastern Cape coast to Cape Town | July to August
Evergreen forest, forest margins, mountain slopes, dune scrub

Crooked trunks tinged yellow-orange; leaves leathery with translucent veins

PLATE 36

299 | *Ekebergia pterophylla* | Rock ash | Rotsessenhout
Mpumalanga, eastern KwaZulu-Natal, north-eastern Eastern Cape | January to March
Rocky outcrops, montane forest, forest margins, ravines

The winged leaf rachis and alternate leaves are characteristic.

PLATE 37

423.1 | *Allophylus africanus* | Black false currant | Swartbastertaaibos
Eastern Mpumalanga, KwaZulu-Natal, north-eastern Eastern Cape |
February to March
Riverine thicket, forest margins, open woodland

Compared to A. decipiens *(Pl. 31), the leaf-blade is noticeably quilted.*

PLATE 38

438 | *Hippobromus pauciflorus* | False horsewood | Basterperdepis
Eastern Limpopo, eastern Mpumalanga, KwaZulu-Natal, eastern Eastern Cape | October to December
Forest margins, rocky ridges, stream banks, bushveld

Slightly winged rachis and unequal leaflets with toothed apices are diagnostic.

PLATE 39

250 | *Erythroxylum pictum* | Forest coca tree | Boskokaboom
Coastal KwaZulu-Natal, coastal north-eastern Eastern Cape | January to April
Forest, forest margins, stream banks, rocky outcrops

The twigs are flattened and the thin leathery leaves are blue-green.

PLATE 40

577 | *Maesa lanceolata* | False assegai | Basterassegai
Eastern Limpopo, eastern Mpumalanga, eastern KwaZulu-Natal | March to December
Evergreen forest margins, stream and river banks, mountain grassland

Small, neat tree; elliptic leaves with prominent veins and serrated margins

PLATE 41

261 | *Vepris lanceolata* [=*Vepris undulata*] | White ironwood | Witysterhout
Widespread in north and eastern South Africa to the Garden Route | May to July
Evergreen forest, riverine bush

Wavy leaflet margins and crushed leaves smelling of citrus are a give-away.

PLATE 42

597 | *Euclea natalensis* | Natal guarri | Natal ghwarrie
Limpopo, Mpumalanga, KwaZulu-Natal, coastal Eastern Cape | October to June
Dune bush, open woodland, river banks, rocky outcrops

Young leaves hairy; can be conspicuously brown

686 | *Tarenna pavettoides* | False bride's bush | Basterbruidsbos
Eastern Mpumalanga, coastal KwaZulu-Natal, north-eastern Eastern Cape |
March to May
Evergreen forest margins, swamp forest

Twigs four-angled; leaves large with pronounced drip-tip; margins noticeably wavy

PLATE 44

667 | *Clerodendrum glabrum* | Tinderwood | Tontelhout
Western North West, Limpopo, Gauteng, Mpumalanga, KwaZulu-Natal, north-eastern Eastern Cape | March to May
Open woodland, forest margins, river banks, coastal dunes

Leaves usually three whorled; smelly when crushed

PLATE 45

700.1 | *Kraussia floribunda* | Rhino coffee | Renosterkoffie
Eastern Mpumalanga, eastern KwaZulu-Natal | February to July
Dune scrub, swamp forest, riverine fringes

Small tree; branches right-angled; flowers and fruit hanging below the twigs

PLATE 46

599 | *Euclea racemosa* | Sea guarri | Seeghwarrie
West coast, Western Cape and Eastern Cape coast | February to May
Coastal dune forest

Limited to sandy coastal fynbos soils; leathery leaf margins tightly rolled under

PLATE 47

(90% of actual size)

325 | *Bridelia mollis* | Velvet sweetberry | Fluweelsoetbessie
Limpopo, North West, Gauteng | January to May
Bushveld, granite koppies

Dense hairy leaves; veins terminate in leaf margins like all the **Bridelia** *spp.*

PLATE 48

717.1 | *Pavetta eylesii* | Broad-leaved bride's bush | Breeblaarbruidsbos
Northern Limpopo, North West | February to June
Bushveld, sandveld bush, stony hills

Small tree with large leaves that can be hairy underneath but hairless above

PLATE 49

505 | *Aphloia theiformis* | Albino berry | Bergperske
Eastern Limpopo, Mpumalanga, northern KwaZulu-Natal | November to January
Evergreen forest, forested ravines, stream banks

A rare tree with angled twigs that zig-zag; the white fruit are unusual.

PLATE 50

723 | *Psychotria capensis* | Black bird-berry | Lemoenbos
Eastern Limpopo, eastern Mpumalanga, coastal KwaZulu-Natal, coastal
Eastern Cape to Knysna | April to August
Evergreen forest, river courses, dune thicket

Small understorey or marginal tree; leaves stiff, glossy; veins prominent below

PLATE 51

578 | *Rapanea melanophloeos* | Cape beech | Kaapse boekenhout
Widespread along eastern South Africa from Limpopo to Cape Town |
September to March
Moist evergreen forest, river fringes

Stiff leaves, petioles purple; often a pioneer forest ecotone species

PLATE 52

(90% of actual size)

717 | *Pavetta edentula* | Large-leaved bride's-bush | Grootblaarbruidsbos
Eastern Limpopo, Mpumalanga, north-eastern KwaZulu-Natal | January to April
Bushveld, rocky hillsides, grassland

Small tree; largest leaves of the **Pavetta** *spp. that are clustered towards the twig ends*

PLATE 53

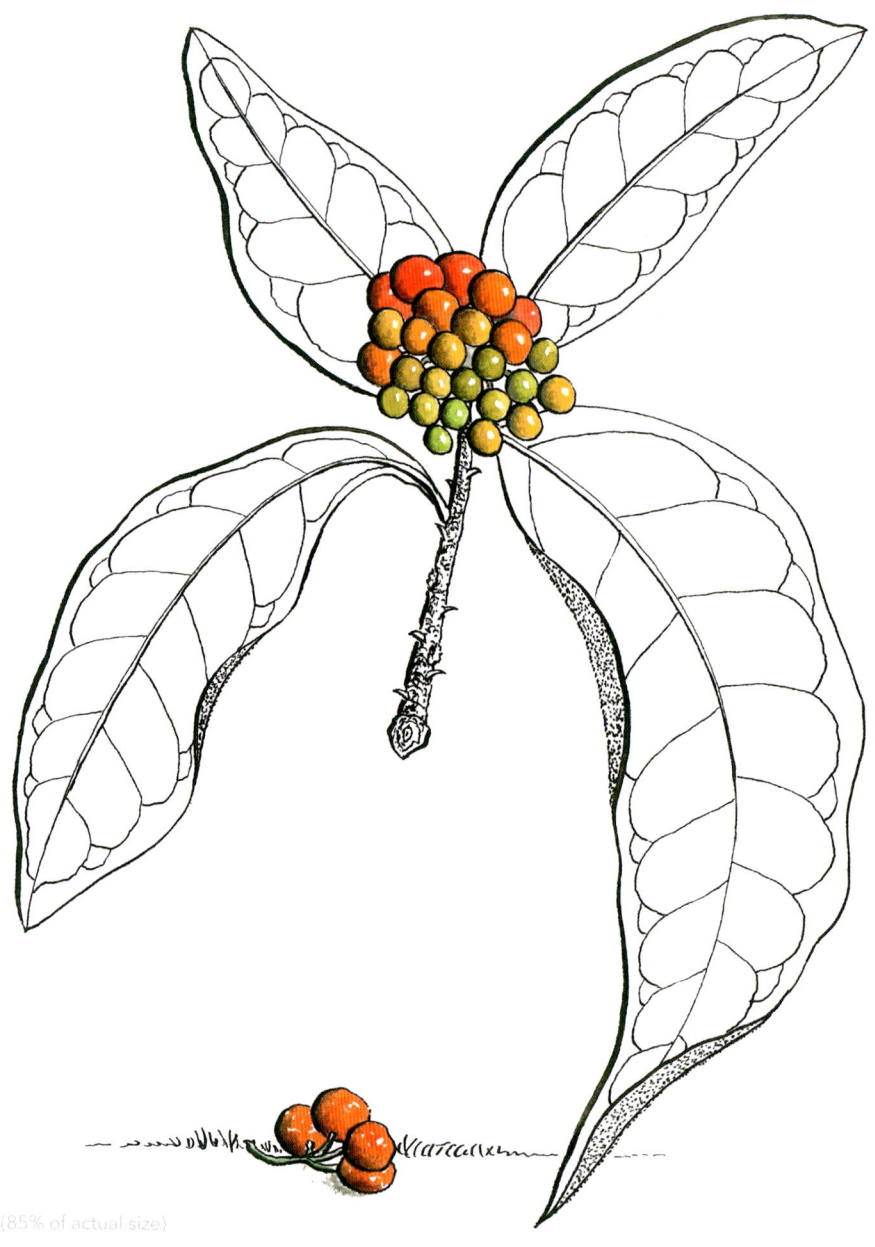

(85% of actual size)

669.4 | *Solanum giganteum* | Red bitter-apple | Grootbitterappel
North West, Limpopo, Gauteng, Mpumalanga, KwaZulu-Natal, coastal Eastern and Western Cape | March to April
Woodland, mountain slopes, forest

Plants can have prickles; thick woolly white hairs on young leaves are diagnostic.

PLATE 54

49 | *Ficus burtt-davyi* | Veld fig | Veldvy
Coastal KwaZulu-Natal, coastal and inland Eastern Cape | August to March
Coastal and mountain forest, dune forest, swamp forest

This is the only scrambling fig.

PLATE 55

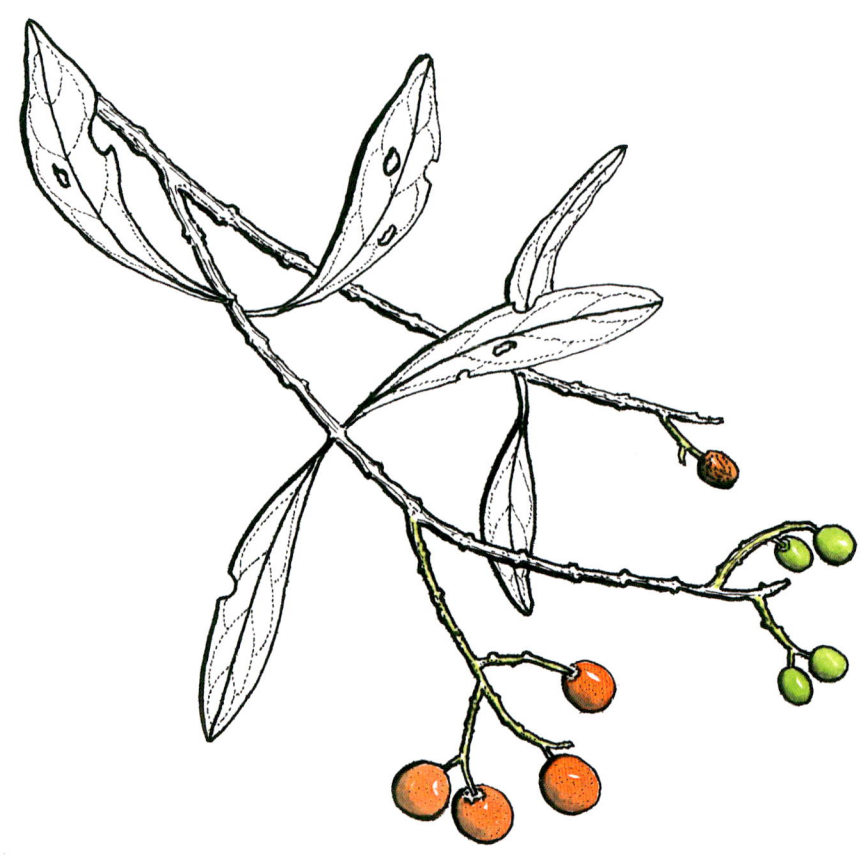

621 | *Salvadora australis* [=*S. angustifolia*] | Narrow-leaved mustard tree | Smalblaarmosterdboom
North-eastern Limpopo, eastern Mpumalanga | November to March
Dry arid bushveld, floodplains

Generally, a dense, rounded shrub or small tree with narrow, greyish leaves

PLATE 56

671 | *Anastrabe integerrima* | Pambati tree | Pambatieboom
Eastern KwaZulu-Natal, north-eastern Eastern Cape | March to July
Evergreen forest, riverine fringes, rocky outcrops

Dark green leaves, usually opposite; creamy brown below with rolled margins

PLATE 57

254 | *Zanthoxylum davyi* [=*Fagara davyi*] | Knobwood | Perdepram
North-eastern Limpopo, Mpumalanga, KwaZulu-Natal, north-eastern Eastern Cape to Garden Route | January to May
Montane forest

Trunks covered with thorn-tipped knobs; prickles also occur on leaf rachis.

PLATE 58

698 | *Tricalysia capensis* | Cape coffee | Kaapse koffie
Eastern Limpopo, eastern Mpumalanga, eastern KwaZulu-Natal, north-eastern Eastern Cape | February to July
Dune forest, forest, riverine fringes

Light brown slightly corky bark; leaves dull-green, petioles hairy; rolled margins

PLATE 59

450 | *Berchemia zeyheri* | Red ivory | Rooi-ivoor
Widespread in Limpopo, Gauteng, Mpumalanga and KwaZulu-Natal |
November to April
Bushveld, open woodland, rocky ridges, water courses

Blocky bark; thin, quilted leaves; petioles and young twigs often a purplish–red

PLATE 60

496 | *Scolopia mundii* | Red pear | Rooipeer
South-eastern Mpumalanga, KwaZulu-Natal, coastal Eastern and Western Cape | October to January
Evergreen forest, forest margins

Leaves stiff and shiny, and margins with hard, sharp, black-tipped serrations

PLATE 61

264 | *Teclea natalensis* | Natal cherry-orange | Natalkersielemon
Eastern Mpumalanga, coastal KwaZulu-Natal, north-eastern Eastern Cape |
November to January
Coastal forests, forest margins, dune thicket

Leaves can be drooping, glossy and aromatic when crushed.

PLATE 62

603.1 | *Diospyros glabra* | Blueberry bush | Bloubessiebos
South-western Western Cape from the coast to the mountains | January to March
Mountain slopes, forest fringes, open grassland, sandy flats

A Western Cape mountain endemic with small leaves that are paler below

PLATE 63

101 | *Olax dissitiflora* | Small sourplum | Bastersuurpruim
Extreme north-eastern Limpopo and eastern Mpumalanga | November to January
Low altitude bushveld, stream banks, rocky outcrops

Slightly drooping, pale green leaves without conspicuous side veins

617 | *Olea europaea* subsp. *africana* [=*Olea africana*] | Wild olive | Olienhout | Currant resin-tree (top); 377 | *Ozoroa sphaerocarpa* | Korenteharpuisboom (bottom)
Widespread (top); Limpopo and Mpumalanga (bottom) | November to June (top); December to January (bottom)
Rocky hillsides, stream banks

O. europaea: *Tough leaves with a herringbone venation and bristle-like apex;* O. sphaerocarpa *leaves roughly hairy and margins finely uneven*

(85% of actual size)

375 | *Ozoroa paniculosa* | Common resin-tree | Gewone Harpuisboom (main illustration); 377 | *O. sphaerocarpa* | Currant resin-tree | Korenteharpuisboom (fruit inset bottom left)
North West, western Limpopo | November to June (main illustration); Northern Limpopo to northern KwaZulu-Natal | December to January
O. paniculosa Bushveld, rocky hillsides; *O. sphaerocarpa* similar habitat and also sandy areas

Both have tough leaves with a herringbone venation and bristle-like apex that are diagnostic. O sphaerocarpa leaf margins are thickened.

PLATE 66

401 | *Maytenus peduncularis* | Cape blackwood | Kaapse swarthout
North-eastern Limpopo, eastern Mpumalanga, KwaZulu-Natal, eastern Eastern Cape, Garden Route | September to November
Evergreen forest, stream banks

Young stems willow-like with soft golden-brown hairs; petioles also hairy

PLATE 67

400 | *Maytenus oleoides* | Rock candlewood | Klipkershout
Western Cape, Garden Route, Little Karoo | December
Rocky mountain slopes

A Cape mountain endemic; bark on old trees thick for protection against fires

PLATE 68

683 | *Crossopteryx febrifuga* | Crystal bark | Sandkroonbessie
Far north-eastern Limpopo | May to August
Dry woodland, sandveld

Glossy, rough-haired leaves with softer hairs below and prominent veins

PLATE 69

398 | *Maytenus acuminata* | Silky bark | Sybas
North-eastern Limpopo, Mpumalanga, KwaZulu-Natal, eastern Free State, Eastern and Western Cape to Cape Town | May to October
Evergreen forest, forest margins, stream banks, montane rocky outcrops

Snapped leaves reveal silky strands that hold the two halves together.

PLATE 70

568 | *Heteromorpha arborescens* [=*H. trifoliata*] | Parsley tree | Wildepietersieliebos
Widespread except in the west | April onwards
Wooded ravines, forest margins, wooded hillsides

Leaves very variable; the peeling bark in node-like bands is diagnostic.

PLATE 71

(85% of actual size)

297 | *Turraea nilotica* | Bushveld honeysuckle tree | Bosveld kamperfoelieboom
Eastern Limpopo, Mpumalanga | September to February
Open woodland, rocky ridges

Northern species, often found in Miombo Woodland, with large, hairy leaves

PLATE 72

399 | *Gymnosporia buxifolia* [=*Maytenus heterophylla*] | Common spikethorn | Gewone pendoring
All of South Africa, except the West Coast | May to January and longer

Leaves variable in shape; the strong, spine-tipped branches are fearsome.

Blueberry size fruit
(Plates 73 to 126)

PLATE 73

601 | *Euclea undulata* | Common guarri | Gewone ghwarrie
Widespread except central Free State and KwaZulu-Natal | April to October
Open woodland, grassland, rocky ridges, semi-desert

A small neat tree with tough leaves and a very undulating margin

PLATE 74

553.2 | *Eugenia natalitia* | Common forest myrtle | Natalmirt
Soutpansberg, Mpumalanga, coastal KwaZulu-Natal, north-eastern
Eastern Cape | November to September
Coastal and montane forest, open woodland

The Eugenia *spp. are not easy to separate vegetatively; fruit can help.*

PLATE 75 & 76

560.1 | *Memecylon bachmannii* [=*M. grandiflorum*] | Pondo rose-apple | Pondoroosappel (top); 560 | *M. natalense* | Natal rose-apple | Natalroosappel (bottom)
Eastern KwaZulu-Natal and Pondoland | April to August (top) & February to August (bottom)
Evergreen and ravine forest, forest margins

These two species are largely distinguished in the field on leaf size.

PLATE 77

660 | *Vitex harveyana* | Three-finger vitex | Kransvingerblaar
South-eastern Mpumalanga, coastal KwaZulu-Natal | March to April
Coastal bush, river valleys, rocky or sandy soils

Most species palmate (has five leaflets) but V. harveyana *usually has three leaflets.*

PLATE 78 & 79

115 | *Cryptocarya myrtifolia* | Myrtle quince | Mirtekweper (top); 117 | *C. wyliei* | Red quince | Rooikweper (bottom)
South-eastern Mpumalanga, coastal KwaZulu-Natal (top); coastal KwaZulu-Natal (bottom) | June to September (top); February to April (bottom)
Evergreen forest (top); Forest margins, rocky outcrops (bottom)

C. myrtifolia: *shiny leaves, pronounced drip-tip;* C. wyliei: *fruit slightly larger, leaf undersurface bluish*

PLATE 80

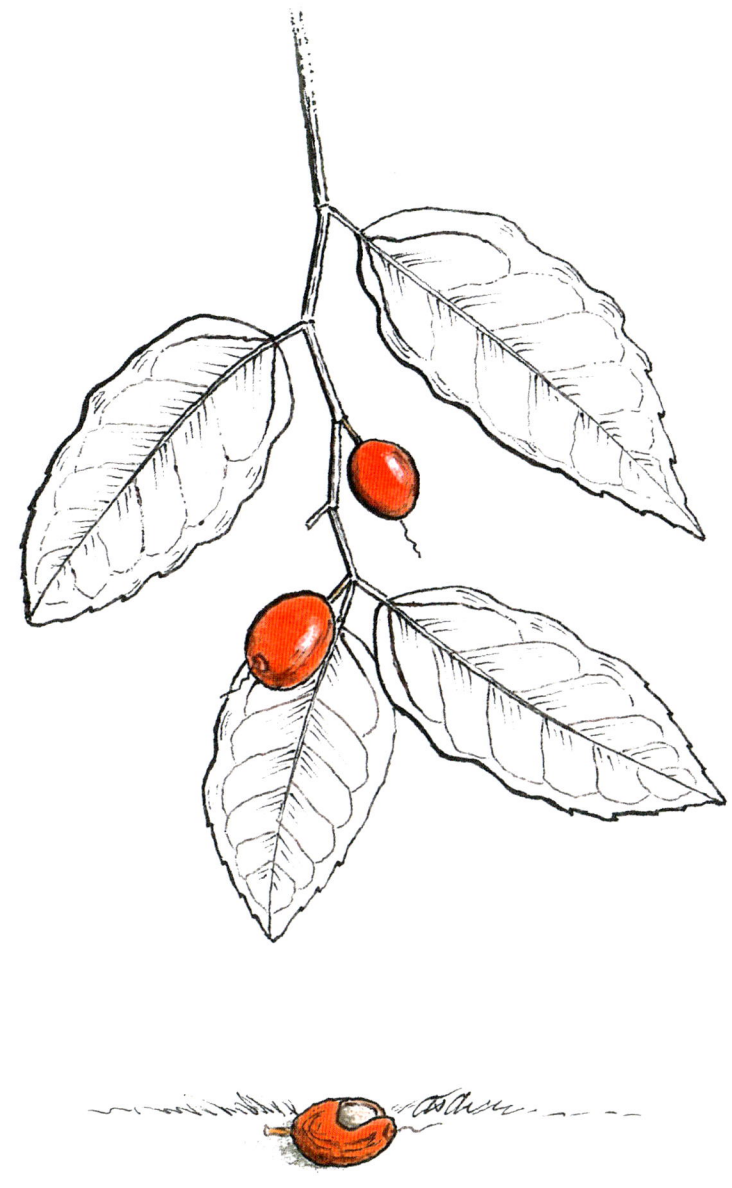

41 | *Celtis mildbraedii* | Natal white stinkwood | Natalse witstinkhout
Isolated patches in Gwalaweni Forest and Durban area | October to November
Sub-tropical coastal forest

Rare large trees with stiff, drooping, shiny leaves and a limited distribution

PLATE 81

418 | *Cassine schinoides* [=*Hartogia schinoides*] | Spoonwood | Lepelhout
Southern regions of Eastern and Western Cape | January to March
Wooded ravines, mountain slopes, stream banks

This Cape forest endemic has narrow, stiff, finely serrated, shiny leaves.

PLATE 82

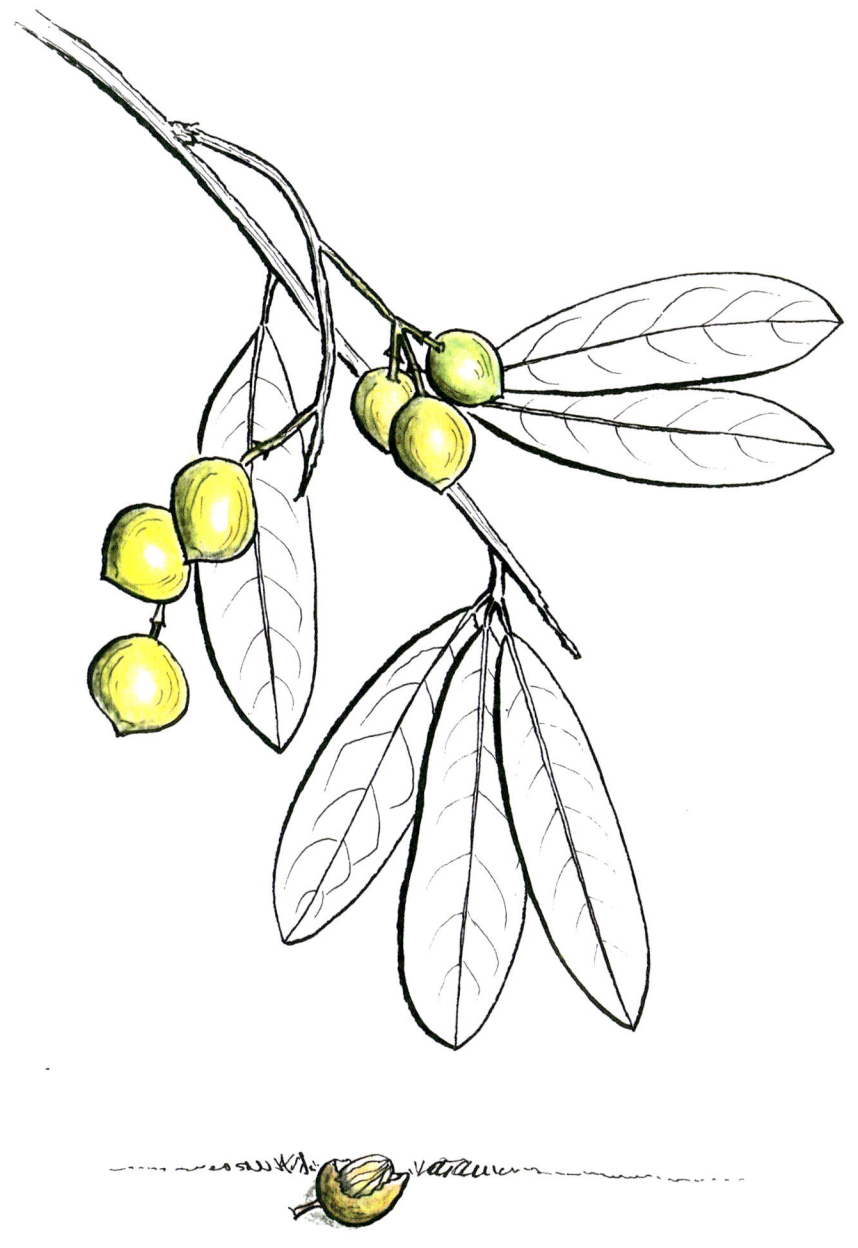

122 | *Boscia albitrunca* | Shepherd's tree | Witgat
Limpopo, central KwaZulu-Natal, western Free State, North West, Northern Cape | October to December
Semi-desert, dry bushveld

White trunk with a dense heavily browsed canopy of small leaves is a give-away.

PLATE 83

(90% of actual size)

55 | *Ficus ingens* | Red-leaved rock fig | Rooiblaarrotsvy
North West, Gauteng, Mpumalanga, KwaZulu-Natal, north-eastern Eastern Cape | June to December
Rocky outcrops, riverine fringes

Unmistakable with clinging white roots and a blaze of bronze spring foliage

PLATE 84

417 | *Maurocenia frangularia* | Hottentot's cherry | Hottentotskersie
Extreme south-western Western Cape | July
Coastal bush, mountain stream banks

A Western Cape coastal endemic; biggish, roundish, very tough leaves with rolled margins

PLATE 85

670 | *Halleria lucida* | Tree fuchsia | Notsung
Western and northern Limpopo, Mpumalanga, KwaZulu-Natal, Eastern and Western Cape coast to Cape Town | August onwards
Evergreen forest, forest margins, karroid scrub

Orange to red tubular flowers, mostly from old wood, is diagnostic.

PLATE 86

570 | *Curtisia dentata* | Assegai | Assegaai
Eastern Limpopo, eastern Mpumalanga, central KwaZulu-Natal, Eastern Cape, Garden Route to Cape Town | May to October
Coastal, montane and scrub-forest

Opposite simple leaves with serrated margins; pink in spring, with rusty hairs

PLATE 87

579 | *Sideroxylon inerme* | White milkwood | Witmelkhout
Coastal KwaZulu-Natal, coastal Eastern and Western Cape | July to January
Coastal woodland, dune thicket, bushveld

Corky-blocky grey bark; milky latex and translucent leaf margins are diagnostic.

PLATE 88

652 | *Cordia caffra* | Septee saucer-berry | Septeeboom
Eastern Limpopo, eastern Mpumalanga, eastern KwaZulu-Natal, coastal Eastern Cape | January
Coastal evergreen forest, riverine bush

Smooth guava tree-like bark and drooping leaves are characteristic.

PLATE 89

627 | *Strychnos mitis* | Yellow bitterberry | Geelbitterbessie
Eastern KwaZulu-Natal, eastern Eastern Cape coastal region | May to July or on to November
Coastal bush and moist evergreen forest

Like all **Strychnos** *spp. leaves are conspicuously three-veined from the base.*

PLATE 90

260 | *Vepris reflexa* | Bushveld white ironwood | Basterwitysterhout
Mpumalanga, north-eastern KwaZulu-Natal | January to May
Rocky hillsides, deciduous woodland, wooded grassland

Reflexed, trifoliolate leaves, often folded upwards are diagnostic.

PLATE 91

414 | *Cassine peragua* | Cape saffron | Bastersaffraan
Along eastern seaboard from KwaZulu-Natal, Eastern and Western Cape | July to August
Evergreen forest, forest margins, dune scrub

Crooked trunks often tinged yellow-orange; leaves thick and leathery

PLATE 92

(85% of actual size)

362 | *Lannea discolor* | Live long | Dikbas
Eastern Limpopo, Mpumalanga, north-eastern KwaZulu-Natal |
September to November
Bushveld and rocky ridges

Pinnately compound bicoloured leaves on thick twigs are characteristic.

PLATE 93

430 | *Deinbollia oblongifolia* | Dune soap-berry | Duineseepbessie
Coastal KwaZulu-Natal, east coastal Eastern Cape | July to October
Dune forest, coastal woodland, riverine fringes

Small, slender, sparsely branched tree that bears fruit above a tuft of leaves

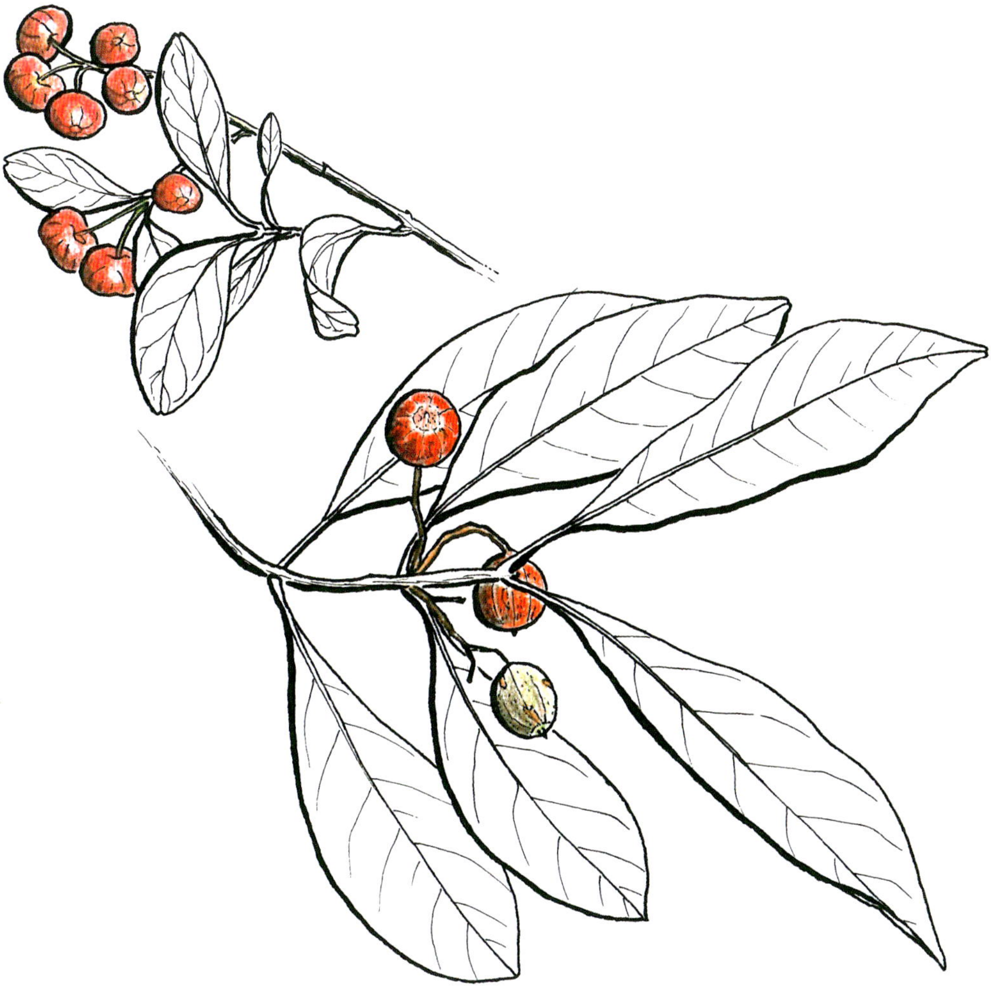

PLATE 94 & 95

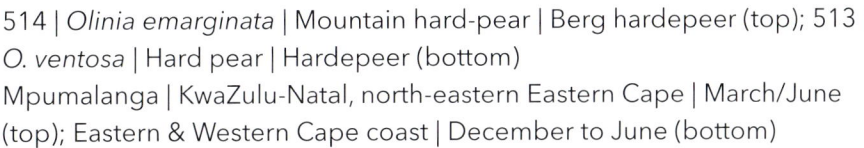

514 | *Olinia emarginata* | Mountain hard-pear | Berg hardepeer (top); 513 | *O. ventosa* | Hard pear | Hardepeer (bottom)
Mpumalanga | KwaZulu-Natal, north-eastern Eastern Cape | March/June (top); Eastern & Western Cape coast | December to June (bottom)

O. emarginata *bark whitish,* O. ventosa *bark darker; crushed leaves smell strongly of almonds.*

PLATE 96

(90% of actual size)

51 | *Ficus cordata* | Namaqua fig | Namakwavy
North-west extremity of Western and Northern Cape | September to February
Arid rocky mountains and river courses

One of two rock-splitting figs of the arid west; leaves ovate on longish petioles

PLATE 97

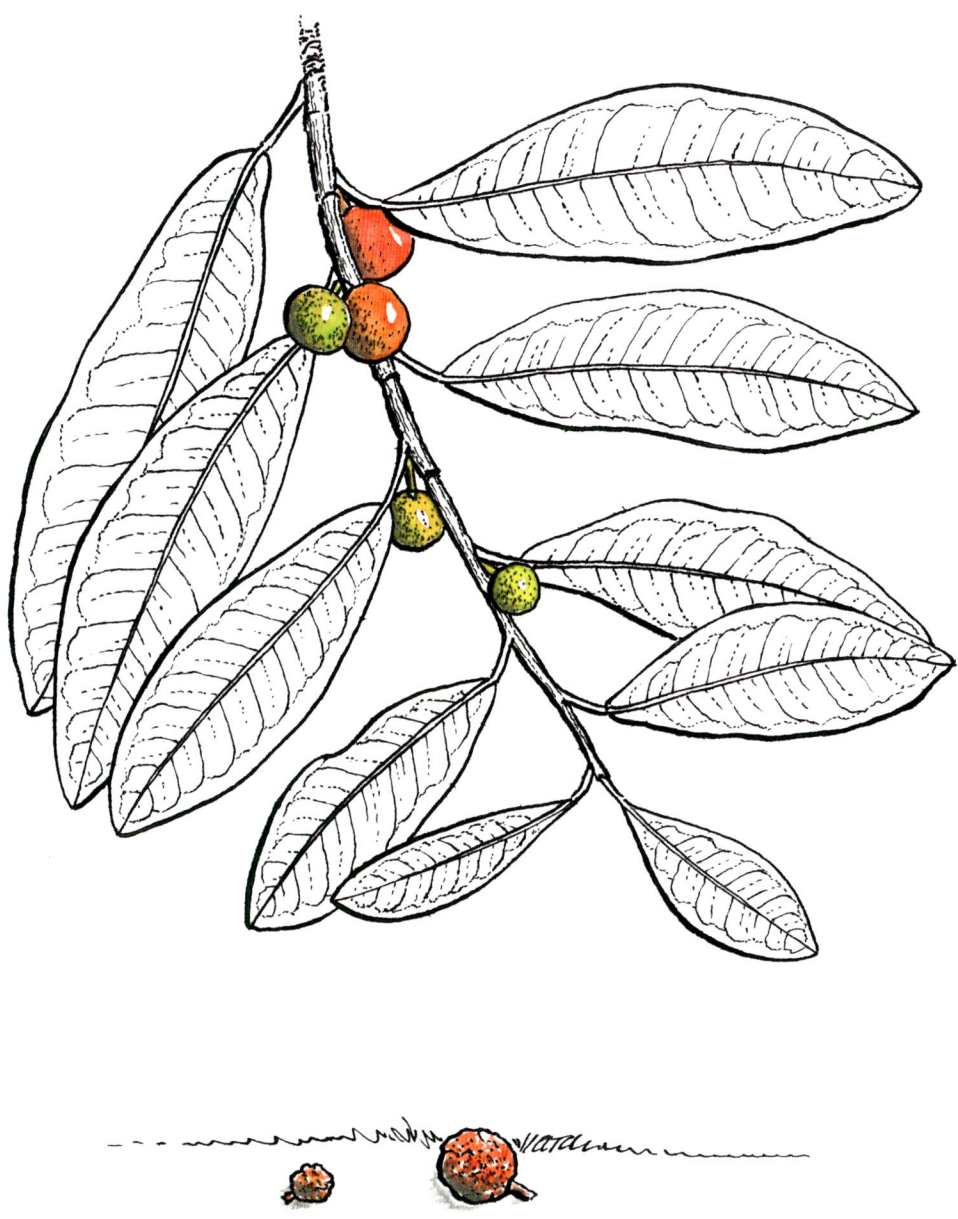

53 | *Ficus ilicina* | Laurel fig | Louriervy
Extreme Western and Northern Cape | September to December
Rocky dolomite and granite outcrops

Another rock-splitting fig with elliptic leaves and much shorter petioles

PLATE 98

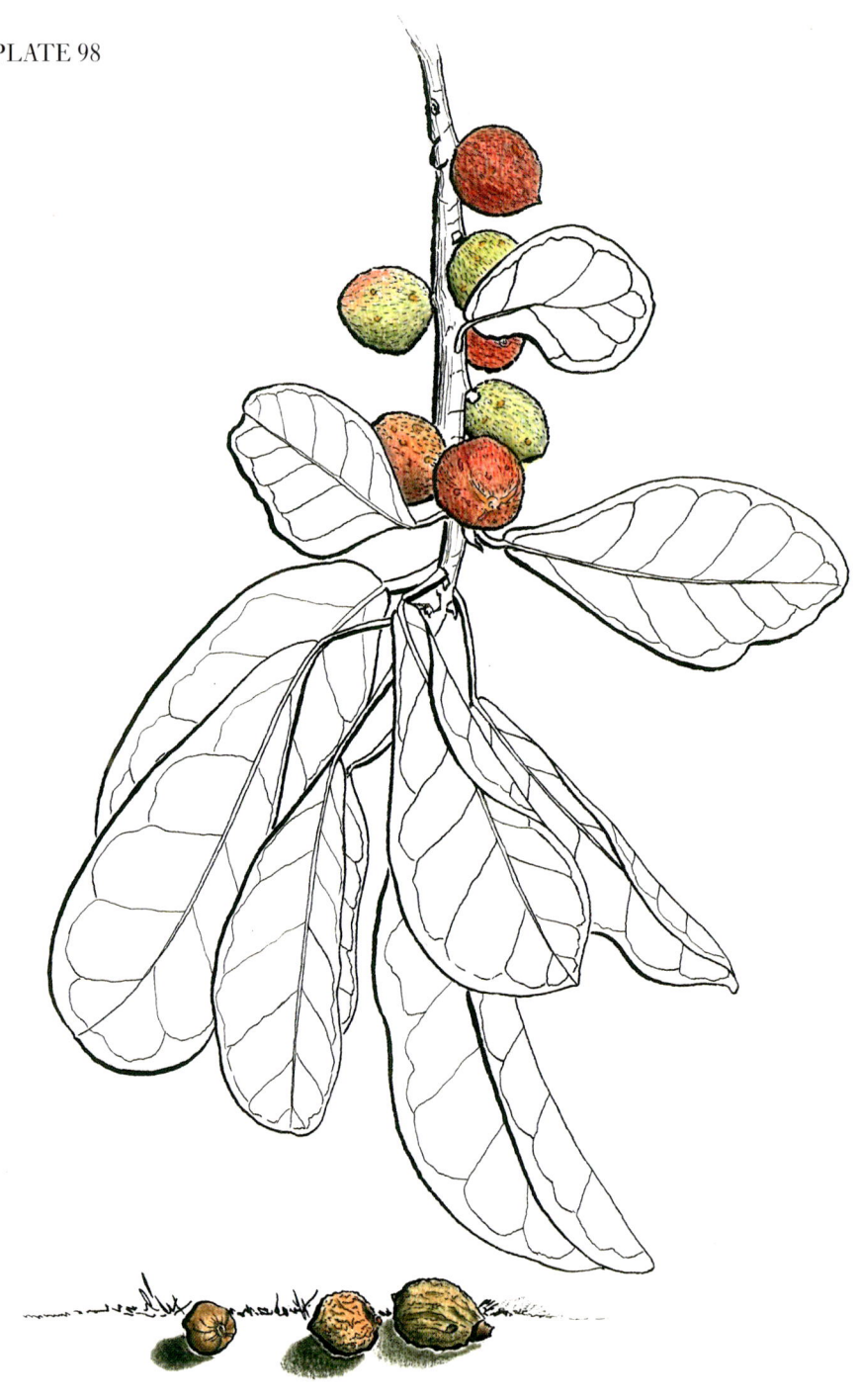

57 | *Ficus natalensis* | Natal fig | Natalvy
Mpumalanga, coastal KwaZulu-Natal, coastal north-eastern Eastern Cape |
July to January
Mesic bushveld and coastal forest

F. natalensis *can be difficult to separate from closely allied species; often a strangler*

PLATE 99

(75% of actual size)

278 | *Commiphora marlothii* | Paperbark corkwood | Papierbaskanniedood
Limpopo and North West | November to March
Arid bushveld, rocky hillsides

Papery bark peels abundantly from a thick succulent trunk.

PLATE 100

351 | *Euphorbia ingens* | Common tree euphorbia or Candelabra tree | Gewone naboom
Limpopo, Mpumalanga, KwaZulu-Natal | August
Deciduous woodland, rocky outcrops, sandy areas

The biggest of our tree Euphorbia *spp. with a dense canopy of angular stems*

PLATE 101

403 | *Maytenus undata* | Koko tree | Kokoboom
Northern Limpopo, central Mpumalanga, KwaZulu-Natal, north-eastern Eastern Cape | March to September
Forest and forest ravines, open woodland, rocky places

New twigs angular, pinkish-purple; leaf venations yellowish below

PLATE 102

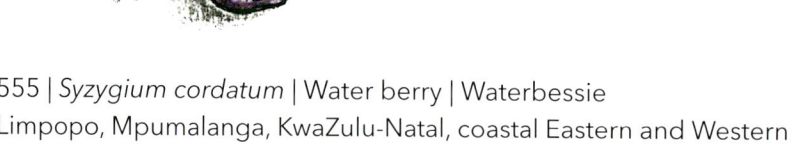

555 | *Syzygium cordatum* | Water berry | Waterbessie
Limpopo, Mpumalanga, KwaZulu-Natal, coastal Eastern and Western Cape | November to March
Riverine thicket, water courses, swamp forest

Branchlets four-angled; leaves mostly with no stalks and bases deeply lobed

PLATE 103

100 | *Osyris quadripartita* [=*O. lanceolata*] | Coastal tannin-bush | Bergbas
Widespread from Limpopo to the Eastern Cape | February and into winter
Rocky ridges, mountain slopes, wooded grassland

Similar to O. compressa *(plate 114) but leaves alternate and not opposite*

PLATE 104

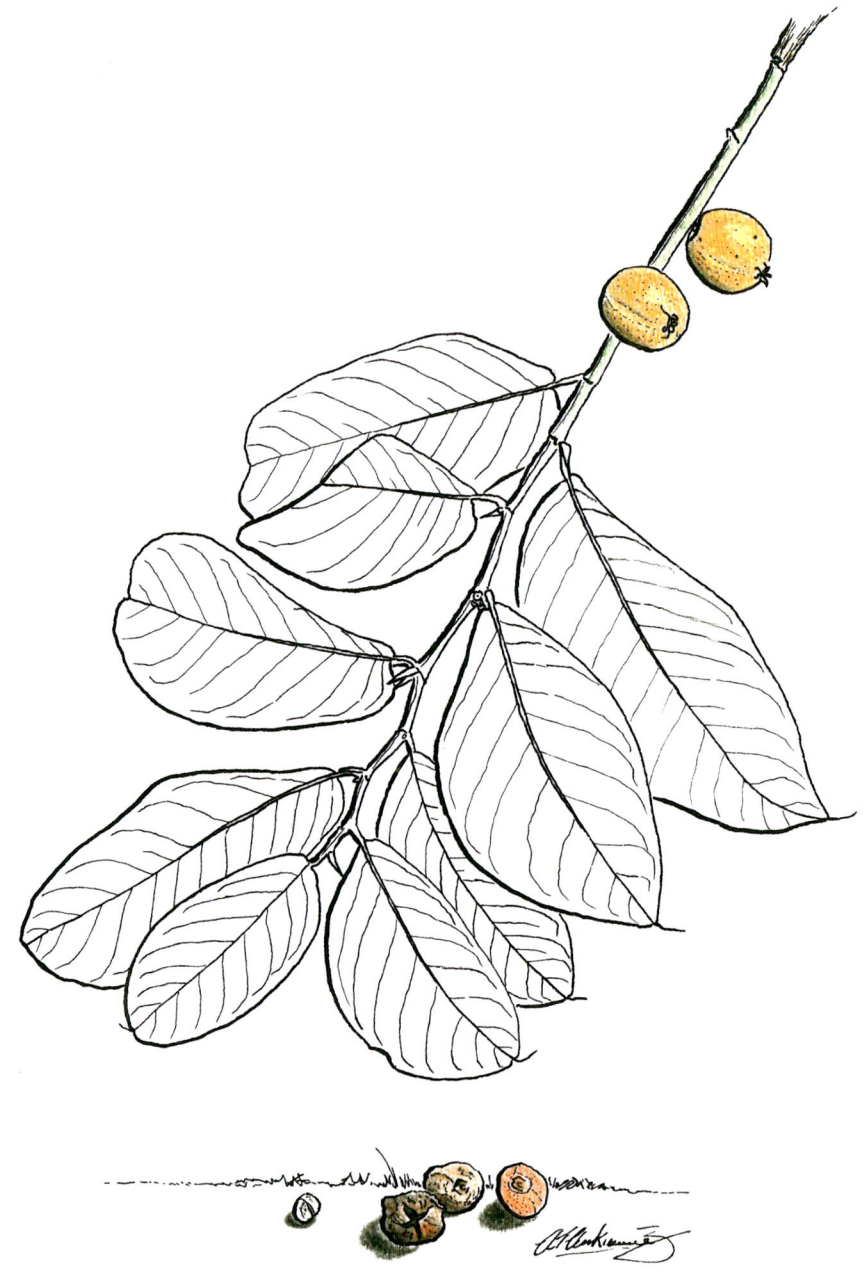

43 | *Chaetacme aristata* | Thorny elm | Doringolm
Gauteng, Limpopo, Mpumalanga, eastern KwaZulu-Natal, north-eastern Eastern Cape | January to September
Water courses, wooded grassland, riverine thicket

Young twigs and coppice zig-zag and leaves end with a sharp bristle-tip.

PLATE 105

285 | *Commiphora glandulosa* [=*C. pyracanthoides*] | Common corkwood | Gewone kanniedood
North West, Limpopo, Gauteng, Mpumalanga | November to February
Dry woodland, Mopane veld

Bark peels in small, yellow, papery flakes and green below; twigs spine tipped

PLATE 106

420 | *Cassinopsis ilicifolia* | Lemon thorn | Lemoendoring
North West, Limpopo, Mpumalanga, western KwaZulu-Natal, Eastern Cape, Garden Route | February to March
Forest margins, riverine forest, stream banks, wooded ravines

Shiny green stems; twigs spine-tipped; leaf tips tapering and curled back

PLATE 107

449 | *Berchemia discolor* | Brown ivory | Bruinivoor
North-eastern Limpopo, Mpumalanga, north-eastern KwaZulu-Natal |
January to May
Bushveld, dry woodland, riverine fringe forest

Bushveld tree with slightly quilted leaves where the side-veins end at the margin

PLATE 108

459.2 | *Grewia flavescens* | Rough-leaved raisin | Skurweblaarrosyntjie
North West, Gauteng, Limpopo, Mpumalanga, north-eastern KwaZulu-Natal | July to August
Bushveld, open woodland, rocky koppies, forest margins

Bushveld shrub or small tree with four-angled twigs; leaves sandpapery

PLATE 109

410 | *Mystroxylon aethiopicum* [=*Cassine aethiopica*] | Kooboo berry | Koeboebessie
North West, Gauteng, Limpopo, Mpumalanga, KwaZulu-Natal, Eastern and Western Cape coast | June to January
Bushveld, evergreen forest, rocky ridges, riverine bush

Trees also occasionally in dry forest; leaves variable

PLATE 110

249 | *Erythroxylum emarginatum* | Common coca tree | Gewone kokaboom
Eastern Limpopo, Mpumalanga, eastern KwaZulu-Natal, eastern Eastern Cape coast | January to May
Evergreen forest understorey, coastal forest fringes and ravines, rocky hillsides

Twigs flattened and leaf apex indented; leaf veins looping towards the margins

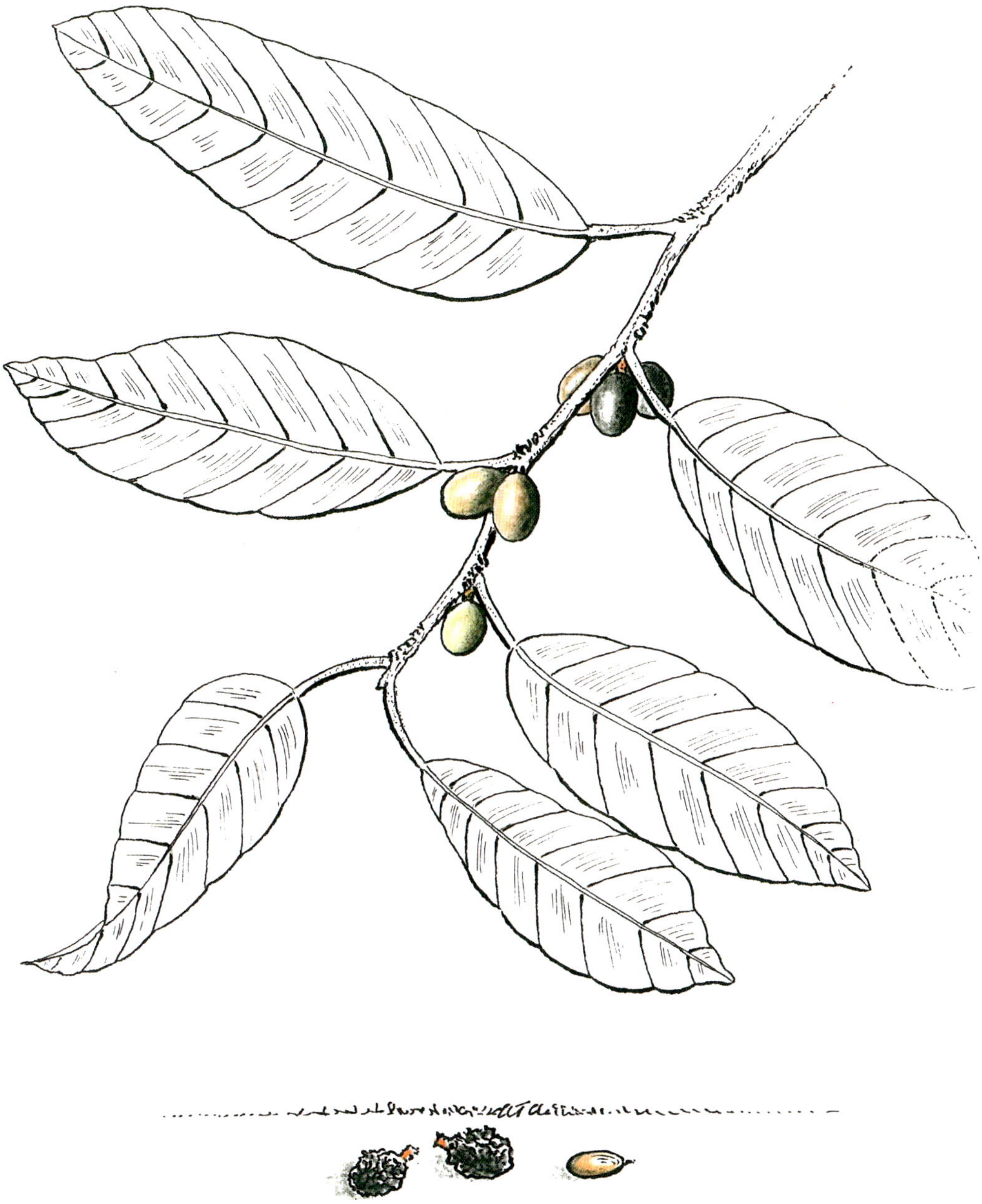

(85% of actual size)

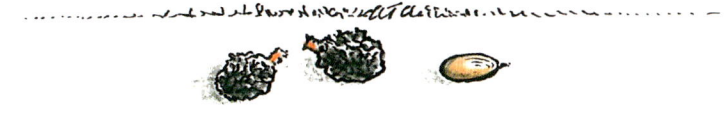

324 | *Bridelia micrantha* | Mitzeeri | Mitseeri
Eastern Limpopo, Mpumalanga, eastern KwaZulu-Natal, north-eastern Eastern Cape | January to March
Riverine forest, coastal swamp forest

Leaf veins end at the margin; twigs slightly zig zagged; spring leaves bronze

PLATE 112

715 | *Canthium armatum* [=*Plectroniella armata*] | False turkey-berry | Basterbokdrol
Limpopo, Mpumalanga, north-eastern KwaZulu-Natal | February to April
Mopane woodland, rocky mountain slopes

Twigs in opposite pairs and can be heavily spined with leaves clustered below

PLATE 113

605 | *Diospyros lycioides* | Bluebush | Bloubos
A wide distribution except south-western areas of Western Cape |
January to May
A wide variety of habitats

A dense bush or small tree sometimes in groves, with mostly hairy young parts

PLATE 114

99 | *Osyris compressa* [=*Colpoon compressum*] | Coastal tannin-bush | Pruimbas
Coastal areas of KwaZulu-Natal, Eastern Cape, Western Cape and south-west Atlantic coast | July to February
Coastal dunes, among rocks and on mountain slopes

A shrub or small bushy tree; blue–green, opposite leaves stiffly upright

PLATE 115

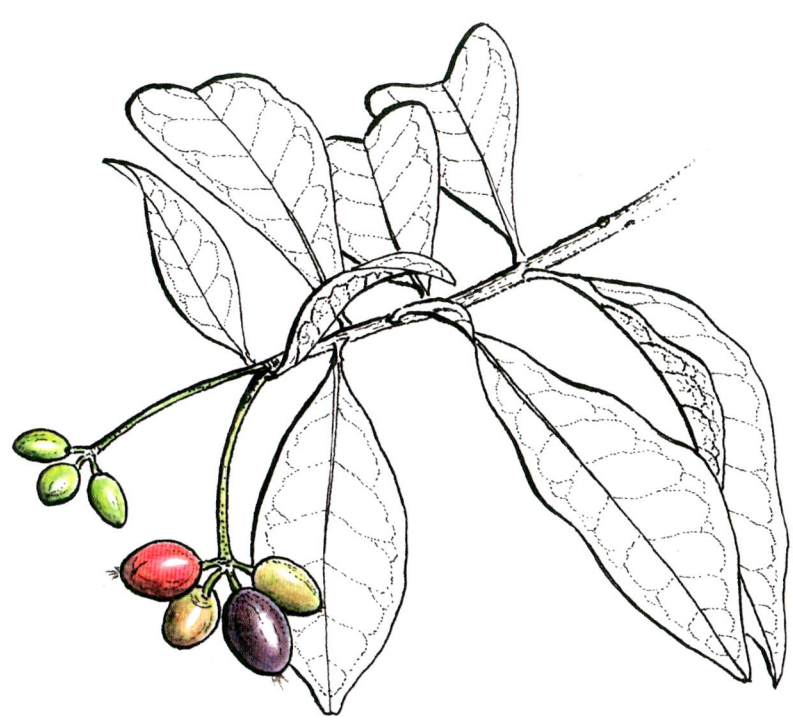

517 | *Peddiea africana* | Poison olive | Gifolyf
Mpumalanga, eastern KwaZulu-Natal, north-eastern Eastern Cape |
February to October
Evergreen forest, forest margins, dense coastal forest behind primary dunes

A shrub or small tree; tough fibrous bark strips off if you try to break a twig.

PLATE 116

592 | *Manilkara discolor* | Forest milkberry | Bosmelkbessie
Coastal KwaZulu-Natal | December to March
Evergreen and sand forest

Can be confused with Englerophytum natalense *(plate 170) with a similar distribution*

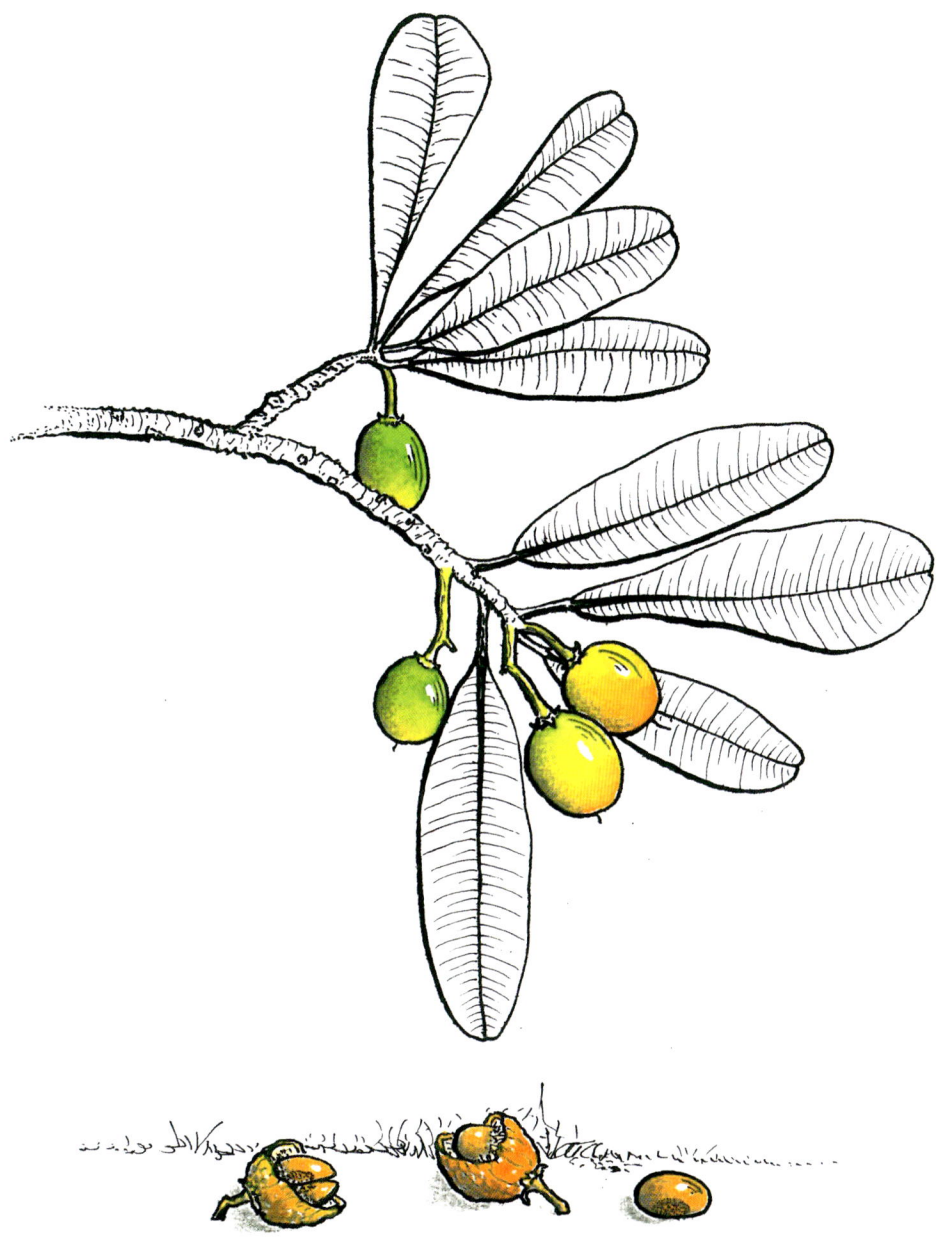

586 | *Manilkara concolor* | Zulu milkberry | Zoeloemelkbessie
South-eastern Mpumalanga, coastal KwaZulu-Natal | October to November
Bushveld, coastal dune forest

Bark rough, grey and longitudinally fissured; leaves ooze a milky sap

PLATE 118

(75% of actual size)

22 | Phoenix reclinata | Wild date palm | Wildedadelpalm
Eastern Mpumalanga, coastal KwaZulu-Natal, Eastern Cape coast |
February to April
River banks, open grassland, coastal dunes

Old stems leaning far over and curving upward again near the end

PLATE 119

(85% of actual size)

364 | *Protorhus longifolia* | Red beech | Rooiboekenhout
Northern Limpopo, central Mpumalanga, coastal KwaZulu-Natal, north-eastern Eastern Cape | October to December
Coastal and montane forest, open woodland, river banks

Occasional bright red to yellow leaves in the canopy are diagnostic.

PLATE 120

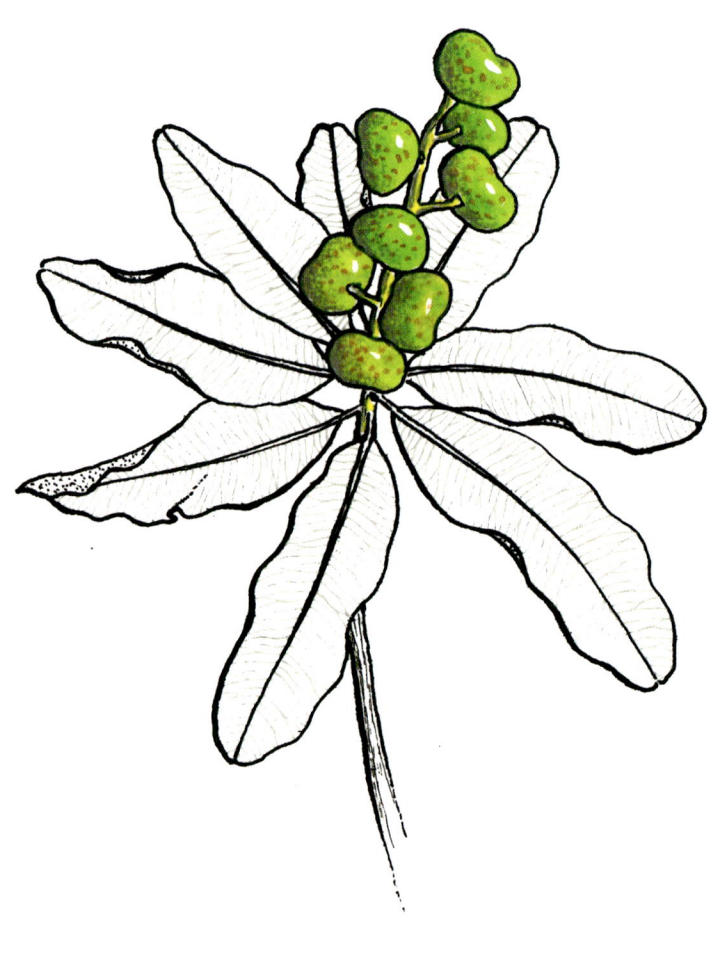

370 | *Ozoroa dispar* | Namaqua resin tree | Namakwaharpuisboom
Western region of Northern Cape | September to November
Semi-arid vegetation, rocky hillsides

Fine herringbone leaf venation and rock-splitting roots are key features.

PLATE 121

276 | *Commiphora glaucescens* | Blue-leaved corkwood | Bloublaarkanniedood
North-west extremity of the Northern Cape | February to April
Rocky hillsides, semi-desert

Yellow to reddish-brown papery bark and clustered leaves are diagnostic.

PLATE 122

263 | *Teclea gerrardii* | Zulu cherry orange | Zoeloekersielemoen
Eastern KwaZulu-Natal, north-eastern Eastern Cape | December to January
Evergreen thicket, dune scrub

Small forest tree; bark flaking in circles; crushed leaves have a citrus-like smell.

PLATE 123

(80% of actual size)

330 | *Croton sylvaticus* | Forest croton | Boskoorsbessie
North-eastern Limpopo, eastern Mpumalanga, eastern KwaZulu-Natal, north-eastern Eastern Cape | March to April
Forest, dense woodland

Like all **Croton** *spp. the two tiny glands at the leaf base are diagnostic.*

PLATE 124

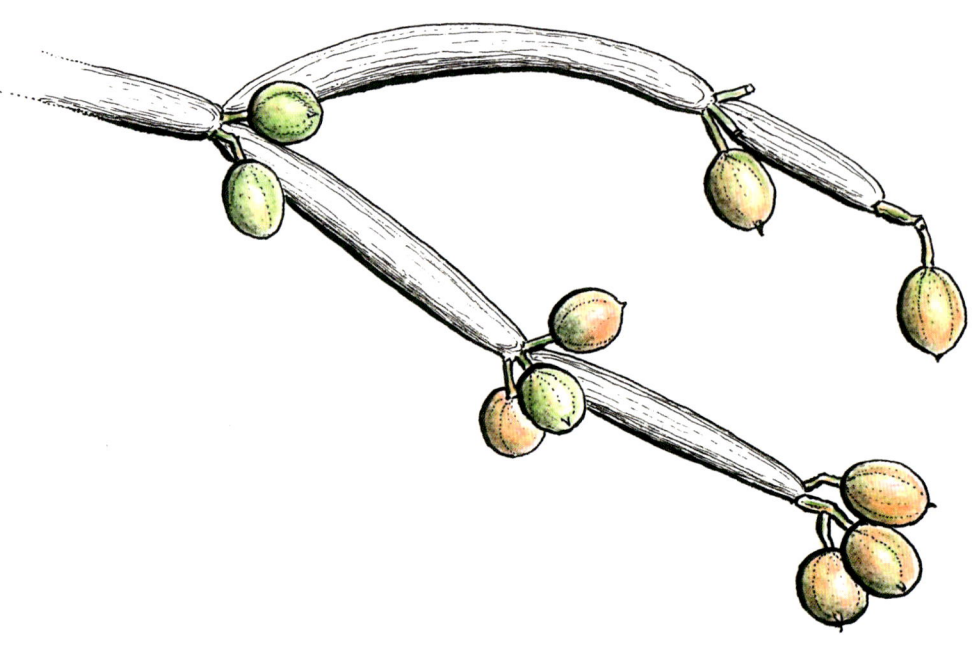

355 | *Euphorbia tirucalli* | Rubber hedge euphorbia | Kraalnaboom
Eastern Limpopo, Mpumalanga, eastern KwaZulu-Natal, north-eastern
Eastern Cape | November to January
Bushveld, rocky slopes, often at old kraal sites

The only large native **Euphorbia** *spp. with rounded stems; the milky latex is very toxic.*

PLATE 125

403.1 | *Putterlickia pyracantha* | False spike thorn | Basterpendoring
South-eastern KwaZulu-Natal, coastal Eastern and West Cape | February to August
River banks, coastal thicket, valley bushveld

Straggling, spiny shrub or small tree; fruit borne on very slender stalks

PLATE 126

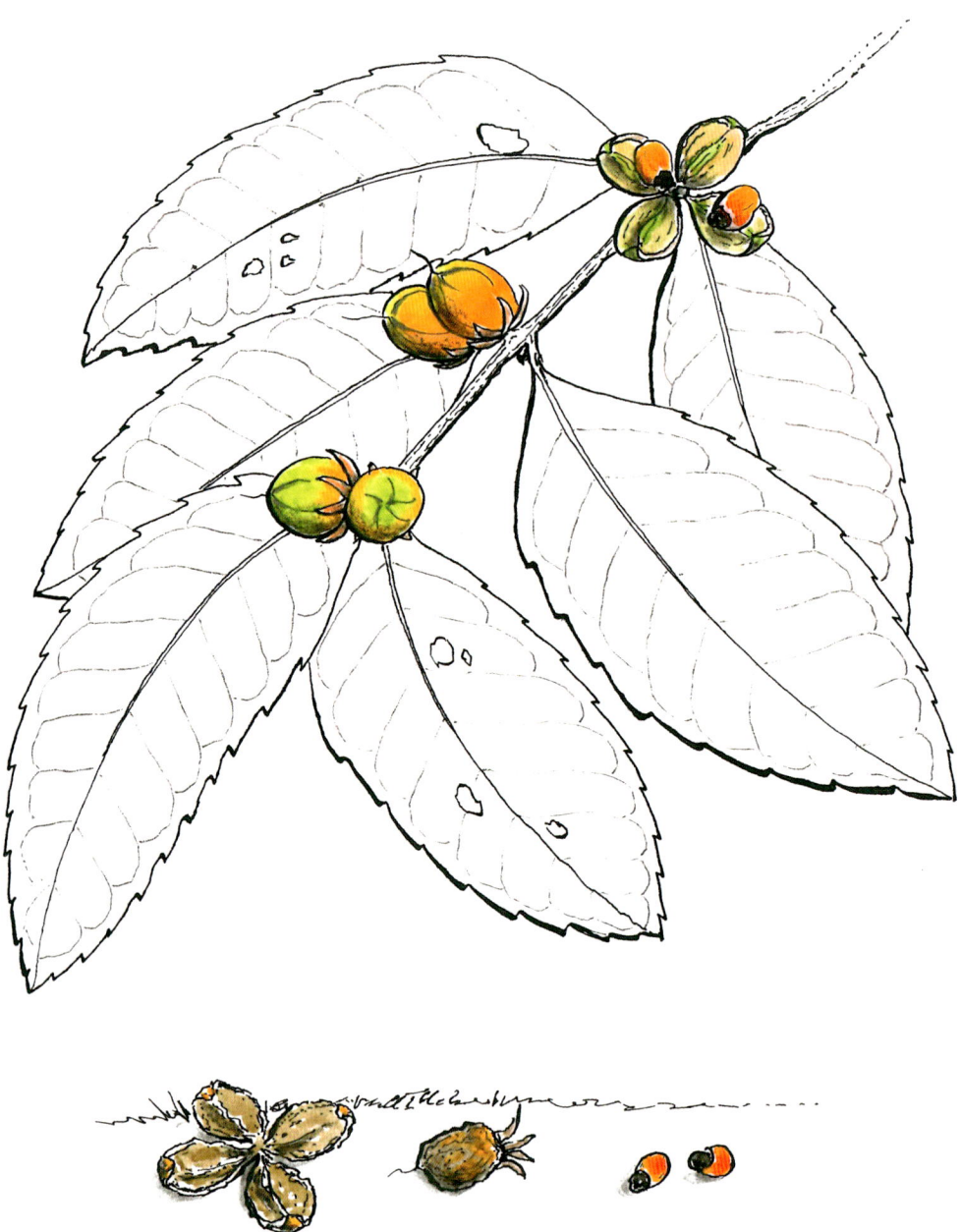

529 | *Cassipourea malosana* [=*C. gerrardii* & *C. congoensis*] | Common onionwood | Gewone uiehout
North-eastern Limpopo, Mpumalanga, north-eastern KwaZulu-Natal | January to May
Evergreen forest, forest margins

Trees with horizontal side branches; terminal buds can be waxy.

Cherry size fruit

(Plates 127 to 172)

PLATE 127

127 | *Boscia mossambicensis* | Broad-leaved shepherds-tree | Breëblaarwitgat
South-eastern Mpumalanga | June to September
Dry woodland, bushveld

A shrub or small tree often with several rigid basal stems; leaves brittle if bent

PLATE 128

465 | *Thespesia acutiloba* | Wild tulip tree | Wildetulpboom
Coastal north-eastern KwaZulu-Natal | February to June
Coastal dune forest and bushveld

Leaves three-lobed and ivy-like with showy hibiscus-like flowers

PLATE 129

703 | *Vangueria parvifolia* [=*Tapiphyllum parvifolium*] | Mountain medlar | Bergmispel
North West, Limpopo, Gauteng, KwaZulu-Natal | January to March
Rocky hills, koppies in high altitude bushveld

Leaves softly hairy on both surfaces; margins slightly wavy and stalks very short

PLATE 130

102 | *Ximenia americana* | Blue sourplum | Blousuurpruim
Limpopo, Mpumalanga, north-eastern KwaZulu-Natal | December to February
Bushveld, hot dry stony slopes

Spinescent bushveld shrub or small tree with blue-green leaves

PLATE 131

553 | *Eugenia zeyheri* | Wild myrtle | Wildemirt
Port Elizabeth to KwaZulu-Natal | January to March
Dry forest margins, riverine thicket, sand forest

In forest situations leaves narrow; margins slightly thickened and rolled under

PLATE 132

631 | *Strychnos usambarensis* | Blue bitterberry | Bloubitterbessie
Limpopo, Gauteng, North West, coastal KwaZulu-Natal, north-eastern Eastern Cape | May to January
Moist evergreen forest, wooded ravines, coastal bush

Opposite leaves three-veined from the base and borne horizontally in one plane

PLATE 133

116 | *Cryptocarya woodii* | Cape quince | Kaapse kweper
Mpumalanga, KwaZulu-Natal, north-eastern Eastern Cape | November to May
Woodland, forest, river valleys, stream banks

Small- to medium-sized tree; upper surface of leaves covered with minute bumps

PLATE 134

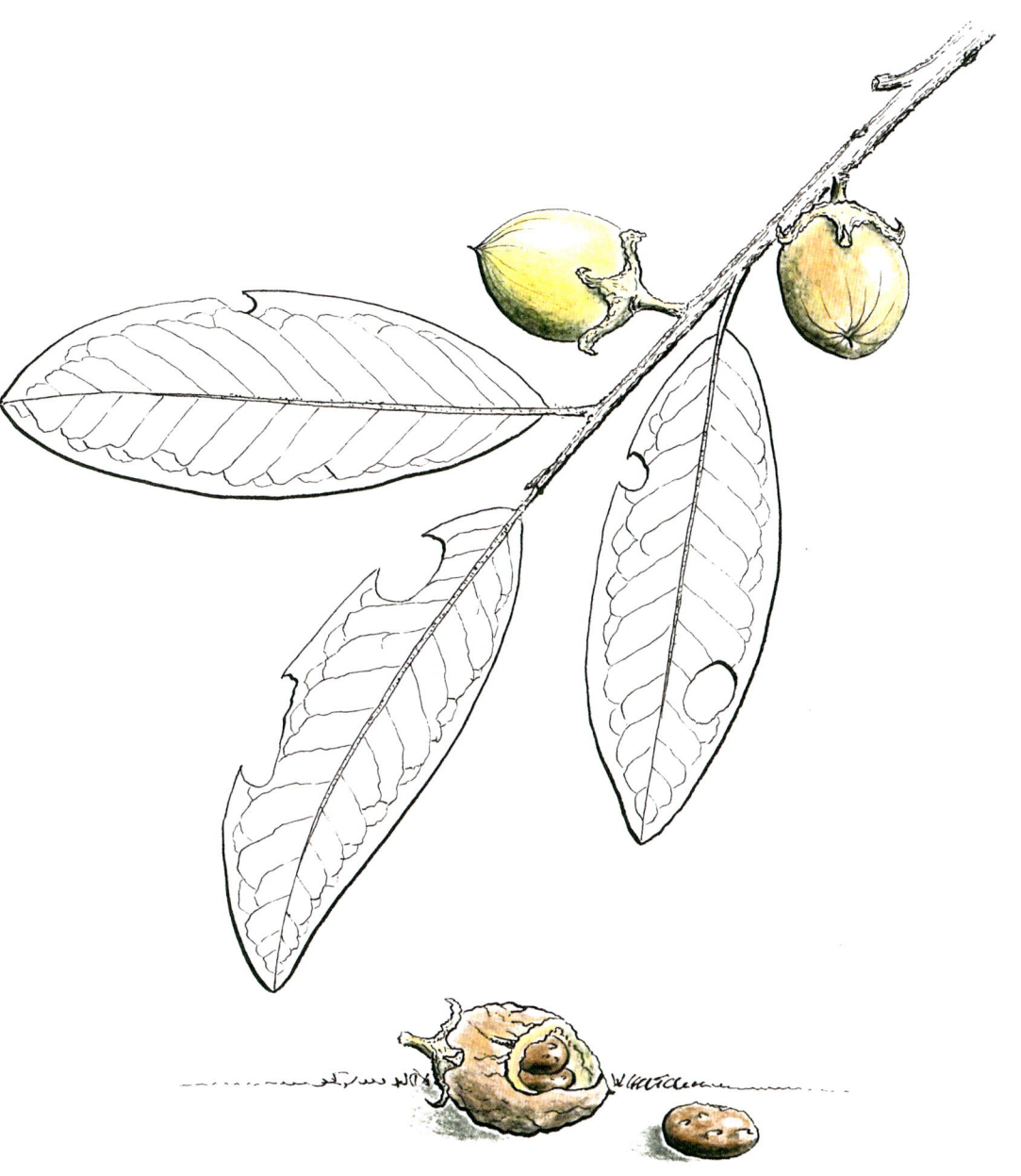

606 | *Diospyros mespiliformis* | Jackal berry | Jakkalsbessie
Eastern Limpopo, Mpumalanga | April to September
Woodland, riverine fringes

Big trees that can be evergreen or deciduous with very dark, straight boles

PLATE 135

(90% of actual size)

17 | *Podocarpus henkelii* | Henkel's yellowwood | Henkel se geelhout
Western KwaZulu-Natal, north-eastern Eastern Cape | September to January
Moist evergreen and montane forest, occasionally in coastal forest

Forest canopy tree with long, pendulous leaves

PLATE 136

447 | *Ziziphus mucronata* | Buffalo thorn | Blinkblaar-wag-'n-bietjie
Widespread except in the Northern Cape and Western Cape | March to August
Bushveld, open woodland, alluvial soils along river courses

The one forward and the other recurved spines diagnostic

PLATE 137

(85% of actual size)

54 | *Ficus trichopoda* [=*F. hippopotami*] | Swamp fig | Moerasvy
Coastal northern KwaZulu-Natal | September to March
Coastal and swamp forest

Lateral branches are supported by huge prop-roots similar to the legs of a hippopotamus.

647 | *Rauvolfia caffra* | Quinine tree | Kinaboom
Eastern Limpopo, Mpumalanga, eastern KwaZulu-Natal, north-eastern Eastern Cape | October to March
Wooded stream banks, forest margins

This forest specialist has whorled, glossy leaves that exude a copious milky latex.

PLATE 139

640 | *Acokanthera rotundata* [=A. schimperi] | Round-leaved poison bush | Rondeblaargifboom
Eastern Limpopo, Mpumalanga | August to November
Bushveld, rocky outcrops

This small tree has round leaves; A. oppositifolia *has longer leaves (see also Plate 156).*

PLATE 140

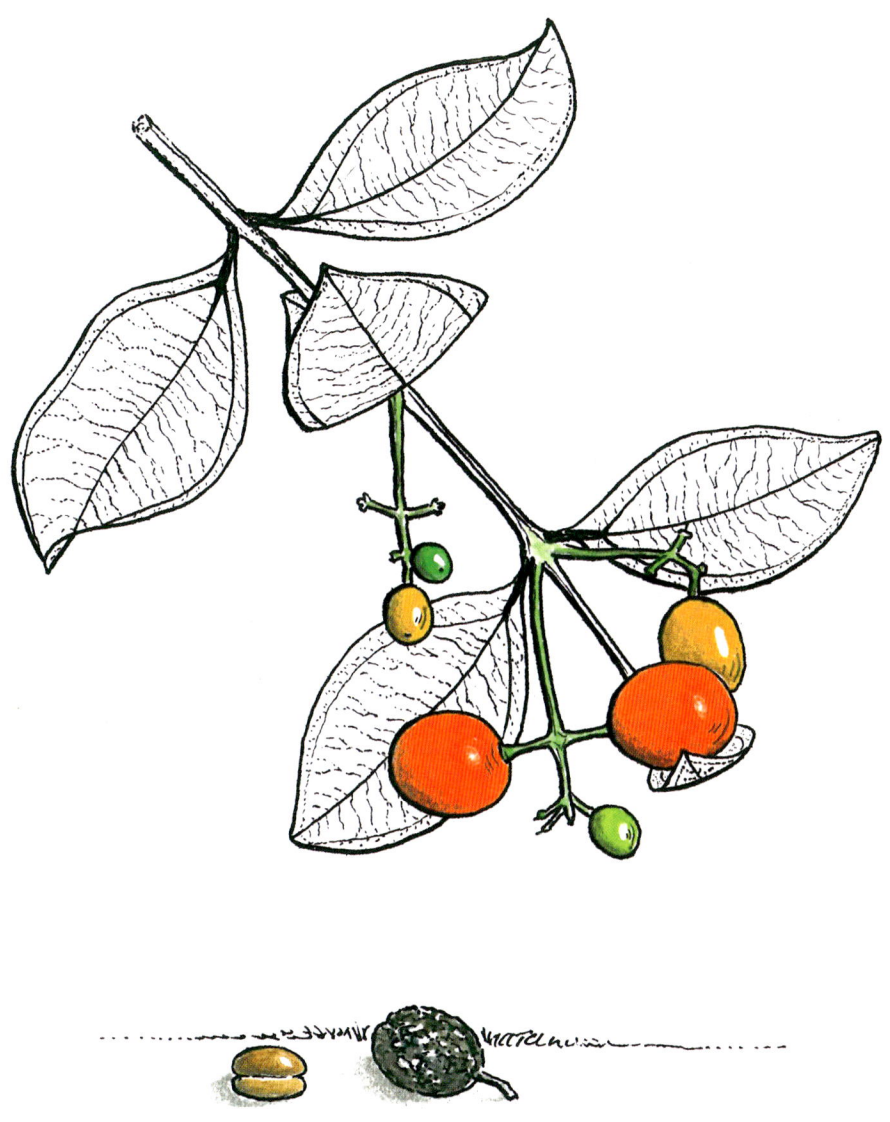

625 | *Strychnos henningsii* | Coffee-bean strychnos | Rooibitterbessie
Mpumalanga, coastal KwaZulu-Natal, north-eastern Eastern Cape |
December to March
Coastal forest, mist belt evergreen forest, riverine fringes

Small tree with yellowish bark; twigs four-angled and opposite; leaves dark green

PLATE 141

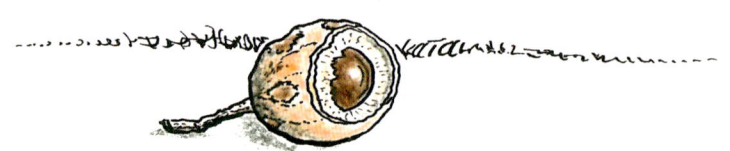

416 | *Elaeodendron transvaalensis* [=*Cassine transvaalensis*] | Bushveld saffron | Bosveldsaffraan
Limpopo, Mpumalanga, eastern KwaZulu-Natal | July to November
Open woodland, bushveld, stream banks

A small tree; arching branches; leaves tightly clustered on the end of short shoots

PLATE 142

654 | *Cordia ovalis* [=*C. monoica*] | Satin-bark saucer-bush | Growweblaarpieringbessie
North-eastern Limpopo, Mpumalanga | December to June
Woodland, bushveld, thicket

Smooth bark and rough sandpaper-like leaves are good clues to this species.

PLATE 143

412 | *Elaeodendron zeyheri* [=*Cassine crocea*] | Small-leaved saffron | Fynblaarsaffraan
North-eastern Limpopo, central KwaZulu-Natal, Eastern and West Cape | March to September
Evergreen forest, forest margins, wooded ravines, rocky outcrops

A small rare tree of rocky outcrops and forests; leaf margins slightly rolled under

PLATE 144

553.1 | *Eugenia capensis* | Dune myrtle | Duinemirt
Coastal KwaZulu-Natal, coastal Eastern Cape to Port Elizabeth | April to December
Pioneer of coastal dunes, occasionally in open woodland and riverine bush

A small tree, very attractive when in flower

PLATE 145

(80% of actual size)

298 | *Ekebergia capensis* | Cape ash | Essenhout
Eastern Limpopo, Mpumalanga, KwaZulu-Natal, north-eastern Eastern Cape |
December to May
Widespread in evergreen and coastal forests

A large tree with green pendulous leaves clustered at the end of the twigs

PLATE 146

624 | *Strychnos decussata* | Cape teak | Kaapse kiaat
North-eastern Limpopo, Mpumalanga, coastal KwaZulu-Natal and Eastern Cape, Garden Route | March to September
Coastal thicket, bushveld, dry water courses

A small slender tree; bark purplish-grey with a fine blocky structure

PLATE 147

(80% of actual size)

278 | *Commiphora marlothii* | Paperbark corkwood | Papierbaskanniedood
North West, Limpopo | November to March
Rocky hillsides, arid bushveld

Green smooth papery bark that peels off in large yellowish pieces

PLATE 148

605 | *Diospyros lycioides* | Bluebush | Bloubos
Widely distribution except west Western Cape | January to May
A wide variety of habitats

Shrubs or small trees often in groves; leaves thin leathery with distinct venation

PLATE 149

262 | *Toddaliopsis bremekampii* | Wild mandarin | Wildenartjie
North-eastern Limpopo, eastern Mpumalanga | February to March
Hot dry woodland, sand forest

A small tree; roundish twigs covered in tiny lenticels; leaflets with short stalks

PLATE 150

669 | *Avicennia marina* | White mangrove | Witseebasboom
Coastal KwaZulu-Natal, coastal north-eastern Eastern Cape | March to September
Estuarine and intertidal areas

The only mangrove with pencil roots that rise out of the tidal mud

PLATE 151

328 | *Croton gratissimus* | Lavender croton | Laventelkoorsbessie
North West, Limpopo, Gauteng, Mpumalanga, north-eastern KwaZulu-Natal | March
Woodland, wooded grassland, rocky outcrops

Like all crotons, the paired basal glands on the leaf base diagnostic, leaves silvery below

PLATE 152

330 | *Croton sylvaticus* | Forest croton | Boskoorsbessie
North-eastern Limpopo, eastern Mpumalanga, eastern KwaZulu-Natal, north-eastern Eastern Cape | March to April
Forest, dense woodland

Like all **Croton** *spp. the two tiny glands at the leaf base are diagnostic.*

PLATE 153

(85% of actual size)

439 | *Bersama lucens* | Glossy Bersama | Blinkblaarwitessenhout
Mpumalanga, eastern KwaZulu-Natal, north-eastern Eastern Cape | August to March
Moist bushveld, forest areas, coastal dunes, rocky places, often in crevices

Small trees marginal to forests or in moist bushveld; new growth reddish–brown

PLATE 154

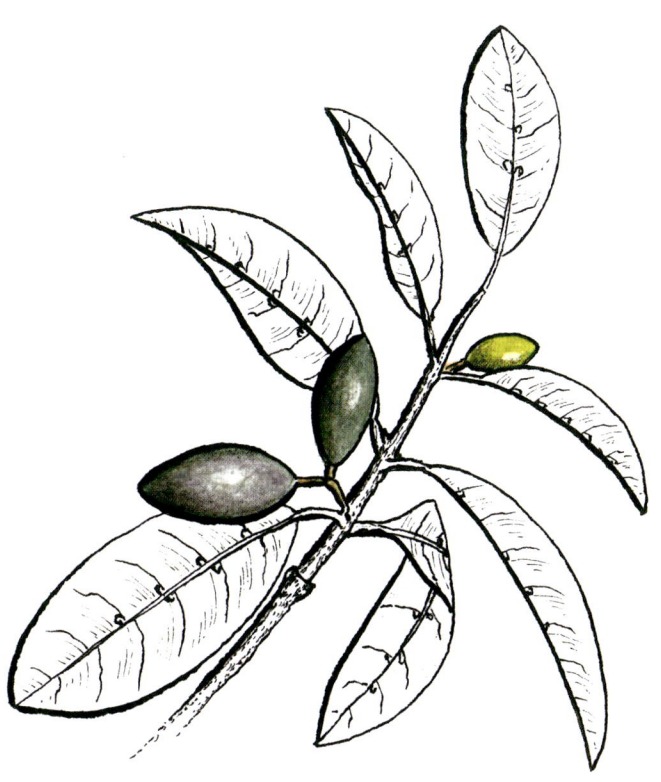

615 | *Chionanthus foveolatus* [=*Linociera foveolata*] | Pock ironwood | Pokysterhout
North West, Limpopo, Mpumalanga, KwaZulu-Natal, Eastern and Western Cape coast | November to July
Montane evergreen forest, wooded ravines, coastal scrub forest

A small tree; horizontal pale twigs; leaves with prominent bumps above

PLATE 155

419 | *Pleurostylia capensis* | Coffee pear | Koffiepeer
Coastal KwaZulu-Natal, eastern Eastern Cape coast | March to June
Coastal and montane forest, wooded ravines, river courses

A rare tree mostly in forest; bark with layers of powdery orange pigment

639 | *Acokanthera oppositifolia* | Common poison bush | Gewone gifboom;
638 | *A. oblongifolia* | Dune poison-bush | Duinegifboom (fruit inset above)
North West, Limpopo, KwaZulu-Natal, Eastern Cape, Garden Route (main illustration); Coastal dunes KwaZulu-Natal to Swellendam (fruit inset) | September to February (main illustration); February to April (fruit inset)
Coastal bush, riverine thicket, open woodland

Both are shrubs or small trees; yellowish latex; all parts of this tree are highly toxic.

PLATE 157

583 | *Mimusops caffra* | Coast red milkwood | Kusrooimelkhout
Coastal KwaZulu-Natal, coastal Eastern Cape | April to September
Coastal dune forest specialist

Leaf apices generally truncated; branchlets covered with brownish matted hairs

PLATE 158

415 | *Elaeodendron croceum* [=*Cassine papillosa*] | Common saffron | Gewone saffraan
Eastern Limpopo, Mpumalanga, eastern KwaZulu-Natal, Eastern and Western Cape coastal regions | January to October
Evergreen forest margins, wooded ravines

The whitish fruit and stiff serrated leaves are diagnostic.

PLATE 159

103 | *Ximenia caffra* | Large sourplum | Grootsuurprium
North West, Limpopo, eastern Mpumalanga, coastal KwaZulu-Natal |
December to January
Woodland, wooded grassland, coastal bush, koppies

Small spinescent tree; leaves clustered on short stalks and folded upwards

PLATE 160

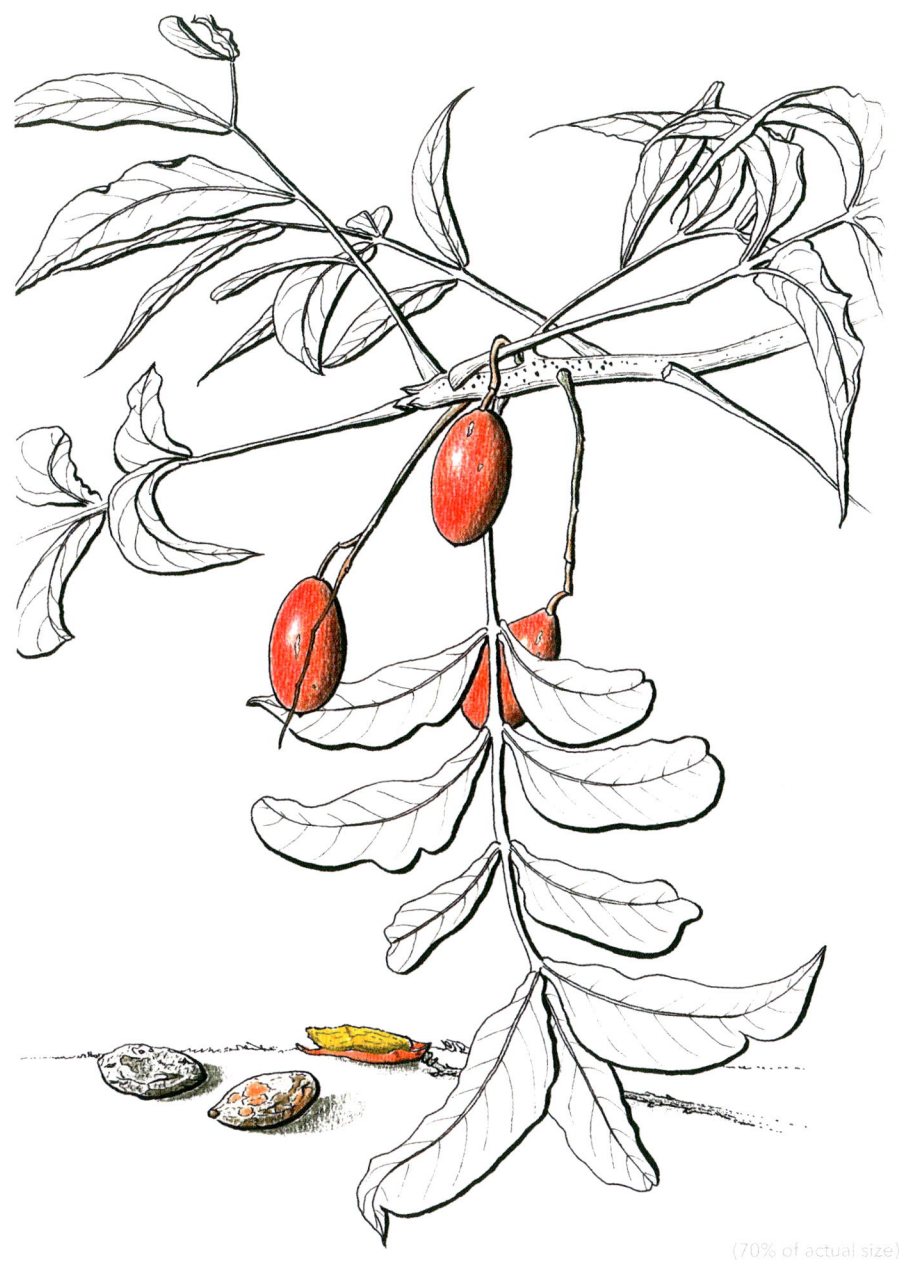

(70% of actual size)

361 | *Harpephyllum caffrum* | Wild plum | Wildepruim
Eastern Mpumalanga, eastern KwaZulu-Natal, north-eastern Eastern Cape |
August to January
Coastal and riverine forest

A large tree with swirled leaves apically clustered on thick twigs

PLATE 161

(90% of actual size)

557 | *Syzygium guineense* | Woodland waterberry | Waterpeer
North-eastern Limpopo, Mpumalanga, north-eastern KwaZulu-Natal |
December to April
Open woodland, vlei margins, mist belt forest, river banks

A big tree with whitish bark and drooping leaves resembling some **Eucalyptus** *spp.*

PLATE 162

(70% of actual size)

275 | *Commiphora edulis* | Rough-leaved corkwood | Skurweblaarkanniedood
Extreme north Limpopo | December to February
Hot dry bushveld, thicket, rocky hillsides

Small tree; bark whitish and flaking in small yellowish papery pieces

PLATE 163

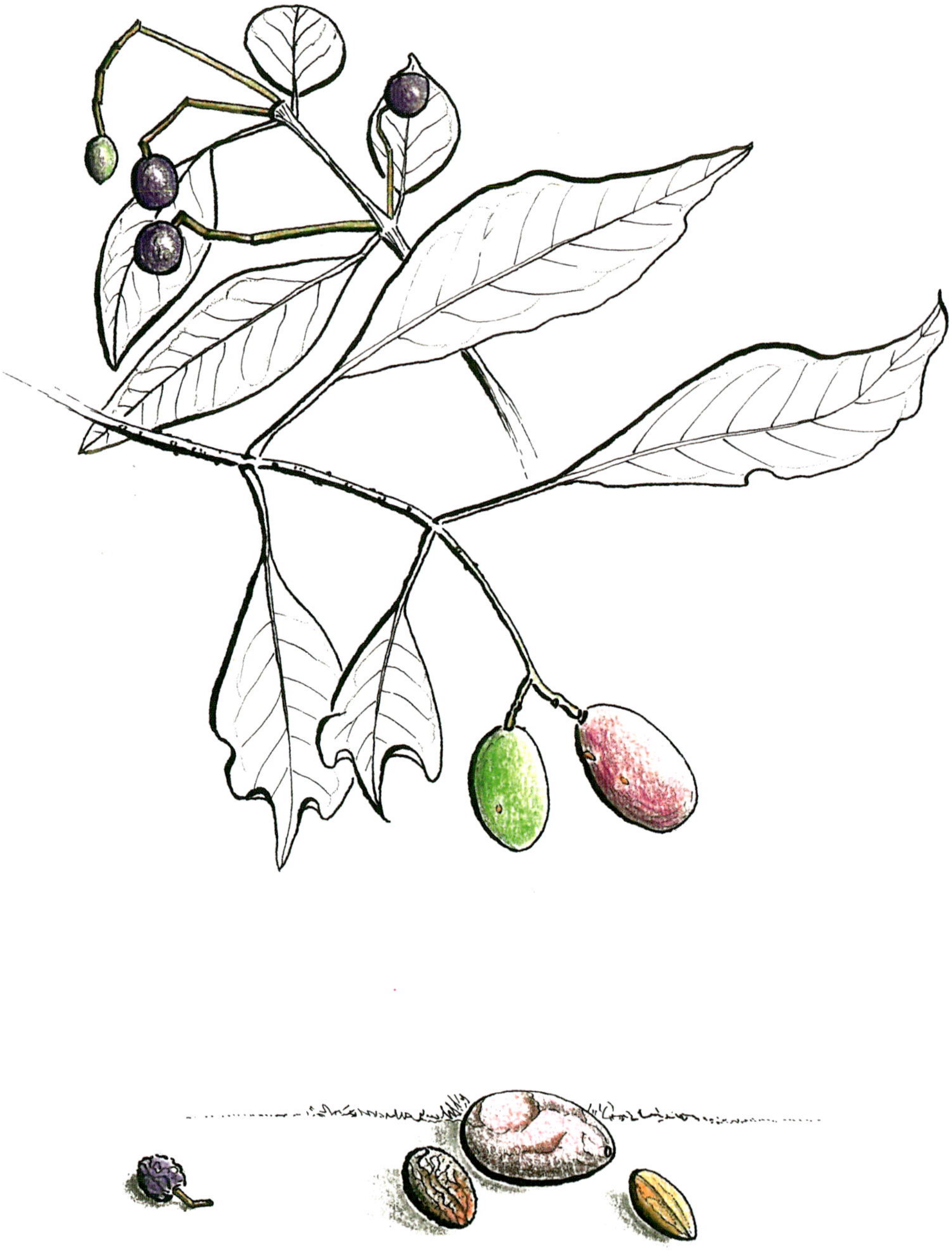

618 | *Olea capensis* subsp. *capensis* | Small ironwood | Kleinysterhout (top);
618.2 | *Olea capensis* subsp. *macrocarpa* | Ironwood | Ysterhout (bottom)
Limpopo, Mpumalanga, KwaZulu-Natal, coastal Eastern and Western Cape | February to September
Evergreen forest, littoral scrub, bush

Medium to large trees; abnormal bark in the form of ring-like crocodile skin

PLATE 164

581 | *Englerophytum magaliesmontanum* [=*Bequaertiodendron magaliesmontanum*] | Stamvrug milkplum | Stamvrug
Eastern North West, Limpopo, Magaliesberg, Mpumalanga, north-eastern KwaZulu-Natal | December to February
Common in rocky outcrops, also in riverine fringes

Generally, a small tree but can be a huge forest giant in Venda

PLATE 165

(85% of actual size)

257 | *Oricia bachmannii* | Twinberry tree | Tweelingbessieboom
Eastern Limpopo, coastal KwaZulu-Natal and north-eastern Eastern Cape coast | April
Montane and coastal evergreen forest

Forest tree with pale trunk and much-branched crown with drooping leaves

PLATE 166

111 | *Xymalos monospora* | Lemonwood | Lemoenhout
Scattered in northern Limpopo, Mpumalanga, KwaZulu-Natal and north-eastern Eastern Cape | November to May
Wet coastal and montane forests

Trunk characterised by bark flaking in concentric markings

PLATE 167

510.1 | *Dovyalis longispina* | Natal apricot | Natalappelkoos (bottom); 509 | *D. rhamnoides* | Common sourberry | Suurbessie (top)
Coastal KwaZulu-Natal and north-eastern Eastern Cape; wider for *D. rhamnoides* | January to February (top); September to November (bottom)
Coastal dune forest, open woodland

D. longispina *a medium-sized tree, pale smooth bark, lenticels on twigs, spines long and thin;* D. rhamnoides *generally a shrub with scalloped leaf margins*

PLATE 168

211 | *Dialium schlechteri* | Zulu podberry | Zoeloepeulbessie
North-eastern KwaZulu-Natal | December to June
Coastal bush, coastal forest, dry sandy soils

Small sand forest tree with a dense canopy of shiny pinnately compound leaves

PLATE 169

511 | *Dovyalis zeyheri* | Wild apricot | Wildeappelkoos
North West, Limpopo, Mpumalanga, KwaZulu-Natal | November to May
Forest margins, open woodland, rocky ridges

A spiny slender tree with shiny leaves that often have a scalloped margin

PLATE 170

582 | *Englerophytum natalense* [=*Bequaertiodendron natalense*] | Silver-leafed milkplum | Natal melkpruim
South-eastern Mpumalanga, eastern KwaZulu-Natal, north-eastern Eastern Cape | September to December
Mixed evergreen and coastal forest

A forest understorey tree in drier forests; leaves clustered at the end of twigs

PLATE 171

359 | *Buxus natalensis* [=*Notobuxus natalensis*] | Natal box | Natalse buksboom
Coastal KwaZulu-Natal and north-eastern Eastern Cape | September to December
Coastal evergreen forest

Small single-stemmed understorey tree usually in groves; corky bark

PLATE 172

433 | *Pappea capensis* | Jacket plum | Doppruim
Widespread except in western KwaZulu-Natal, Free State and south-western Western Cape | February to July
Woodland, riverine bush, karroid vegetation

A widespread bushveld tree; leaves variable; yellowish herringbone leaf veins

175

Table grape size fruit

(Plates 173 to 206)

PLATE 173

308 | *Pseudolachnostylis maprouneifolia* | Kudu berry | Koedoebessie
Northern Limpopo | May onwards
Bushveld, wooded grassland, rocky outcrops

A deciduous tree; old and very young fruit often on the tree together

PLATE 174

113 | *Cryptocarya latifolia* | Broad-leaved quince, Breëblaarkweper
Coastal KwaZulu-Natal | January to March
Evergreen forest, riverine fringes

Leaves leathery with three veins from the base and parallel to the leaf margin

PLATE 175

485 | *Garcinia gerrardii* | Forest mangosteen | Bosgeelmelkhout
Mpumalanga, eastern KwaZulu-Natal, north-eastern Eastern Cape | March to June
Coastal and montane forest

A forest margin tree; foliage dark green; young leaves conspicuously pale green

PLATE 176

663 | *Vitex mombassae* | Poora berry | Poerabessie
Northern Limpopo | January to June
Bushveld, rocky outcrops, deciduous woodland, especially on Kalahari sands

Small deciduous tree; stalkless leaflets can be three or five foliate.

PLATE 177

16 | *Afrocarpus falcatus* [=*Podocarpus falcatus*] | Outeniqua yellowwood | Outeniekwageelhout
Eastern Mpumalanga, KwaZulu-Natal, Eastern Cape and West Cape coast to Swellendam | September to May
High moist forest, wooded ravines

Largest of our forest trees; grey flaky bark and small, sickle-shaped leaves

PLATE 178

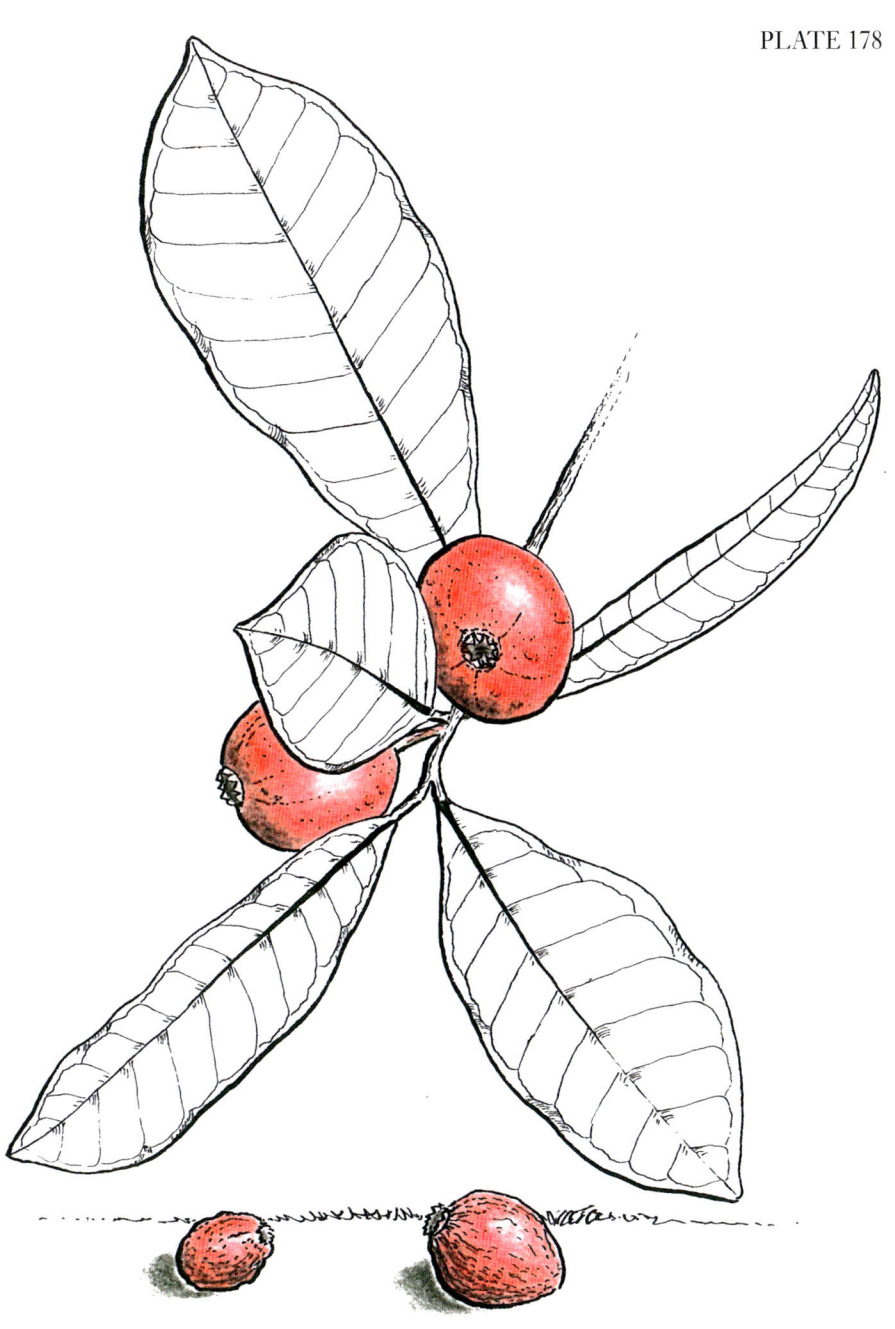

553.3 | *Eugenia erythrophylla* | Large-leaved myrtle | Grootblaarmirt
Coastal KwaZulu-Natal | November to March
Coastal and montane forest

A scarp forest specialist; large leathery leaves that are pinkish-red when young

PLATE 179

(80% of actual size)

702 | *Vangueria infausta* | Wild medlar | Wildemispel
North West, Limpopo, Gauteng, Mpumalanga, eastern KwaZulu-Natal |
January to April
Wooded grassland, bushveld, coastal forest and scrub, rocky places

A small, large-leaved deciduous tree; leaves velvety when young

PLATE 180

(75% of actual size)

63 | *Ficus abutilifolia* [=*F. soldanella*] | Large-leaved rock-fig | Grootblaarrotsvy
North West, Limpopo, Mpumalanga, KwaZulu-Natal | August to February
Bushveld, stream banks, rocky outcrops

The large, heart-shaped leaves and white rock-clinging roots are diagnostic.

PLATE 181

702.1 | *Vangueria cyanescens* | Bush medlar | Blinkblaarmispel
North-eastern Limpopo, Mpumalanga, coastal KwaZulu-Natal | December to February
Bushveld, evergreen dune forest

A small deciduous tree with large thin leaves that can also be hairless

PLATE 182

585 | *Mimusops zeyheri* | Red milkwood | Moepel
North West, Gauteng, Limpopo, Mpumalanga | April to October
Rocky hillsides, forest margins, riverine fringes

A dry woodland tree with a heavy dark green canopy; young leaves hairy

PLATE 183

603 | *Diospyros dichrophylla* | Common star apple | Gewone sterappel
Eastern KwaZulu-Natal, eastern Eastern Cape, Garden Route, Little Karoo, Overberg | April to October
Coastal scrub, open grassland, wooded ravines, wooded rocky hillsides

A small tree with ascending branches; young parts with yellow velvety hairs

PLATE 184

66 | *Ficus sycomorus* | Sycomore fig | Gewone trosvy
Northern Limpopo, Mpumalanga, north-eastern KwaZulu-Natal | July to December and longer
Riverine thicket, mixed woodland

Big trees mostly with figs on the trunk and branches; leaves sand–papery

PLATE 185

(85% of actual size)

316 | *Drypetes natalensis* | Natal ironplum | Natalysterpruim
Coastal KwaZulu-Natal | December to March
Dune and inland forest

Forest understorey tree; fruit on old wood; leaf margins can be spiky like Holly

PLATE 186

507 | *Dovyalis caffra* | Kei-apple | Keiappel
North-eastern Limpopo, Mpumalanga, eastern KwaZulu-Natal, eastern Eastern Cape | November to January
Coastal forest, wooded grassland, riverine thicket

Small spiny tree; usually multi-stemmed; leaves clustered on old wood

PLATE 187

368 | Heeria argentea | Rockwood | Kliphout
Inland south-west of Western Cape | September to January
Dry rocky mountain slopes

A small tree with a dense canopy, restricted to the Cape fold mountains

PLATE 188

695 | *Rothmannia globosa* | Bell gardenia | Klokkieskatjiepiering
Mpumalanga, eastern KwaZulu-Natal, north-eastern Eastern Cape | January to March
Dune thicket, evergreen forest margins, coastal woodland, riverine fringes

A slender tree; leaves have a reddish mid-rib; hairy pockets in the vein axils.

PLATE 189

332 | *Cavacoa aurea* | Natal hickory | Natalokkerneut
Coastal KwaZulu-Natal | December to February
Evergreen forest

A rare forest tree; trunk conspicuously fluted; glossy leaves, deeply channeled leaf stalk

PLATE 190

319 | *Hyaenanche globosa* | Hyaena-poison | Wolwegifboom
North-western Western Cape (Gifberg) | October to December
Arid mountain fynbos

A small tree only on the Gifberg, West Cape; leaves stiff, dark green, four-whorled

PLATE 191

296 | *Turraea floribunda* | Wild honeysuckle tree | Wildekamperfoelieboom
Coastal KwaZulu-Natal, north-eastern Eastern Cape | February to July
Open woodland, rocky places, ravine forest

A small deciduous tree; often a scrambler; branches spreading horizontally

PLATE 192

494 | *Kiggelaria africana* | Wild peach | Wildeperske
Widespread except in the drier western areas of South Africa | February to July
Evergreen forest, rocky outcrops, mountain grassland, wooded ravines

In wind the bicoloured leaves are conspicuous and can be highly variable in shape and form.

PLATE 193

(80% of actual size)

466 | *Azanza garckeana* | Snot apple | Snotappel
Northern Limpopo | February to September
Bushveld

A small tree; bears yellow hibiscus-like flowers with a maroon centre patch

PLATE 194

252 | *Balanites pedicellaris* | Small torchwood | Kleingroendoring
Northern coastal KwaZulu-Natal | May to June
Dry bushveld, alluvial floodplains

Small spiny bushveld tree with grey-green, leathery and almost succulent leaves

PLATE 195

(80% of actual size)

241 | *Xanthocercis zambesiaca* | Nyala tree | Njalaboom
Northern Limpopo, north-eastern Mpumalanga | November onwards
Hot dry bushveld, rich alluvial soils in river valleys

Large trees usually with heavily browsed coppice shoots surrounding the trunk

PLATE 196

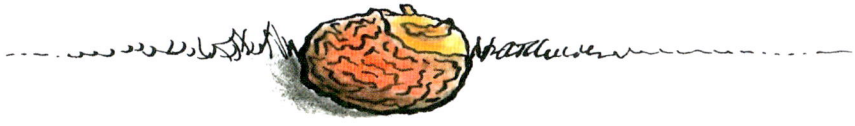

103 | *Ximenia caffra* | Large sourplum | Grootsuurpruim
North West, Limpopo, eastern Mpumalanga, coastal KwaZulu-Natal |
December, January
Woodland, wooded grassland, rocky hillsides, coastal bush

Small spinescent tree; leaves clustered on short stalks and folded upward

(90% of actual size)

696.2 | *Oxyanthus pyriformis* [=*O. natalensis*] | Natal loquat | Natallukwart
Eastern KwaZulu-Natal | February to April or later
Moist evergreen forest

Small rare understorey forest tree; leaves large, with rounded asymmetric base

PLATE 198

(90% of actual size)

251 | *Balanites maughamii* | Torchwood | Groendoring
Eastern Limpopo, Mpumalanga | November to January
Dry bushveld, sand forest, river banks, around pans

Large bushveld tree with a conspicuously fluted trunk and stout branched spines

PLATE 199

640.1 | *Carissa macrocarpa* | Large num-num | Grootnoemnoem
Coastal KwaZulu-Natal, north-eastern Eastern Cape coast | September to January
Coastal scrub and sand dunes

A small lowland shrub, rarely a small tree; leaves glossy; spines branched and a copious milky latex

PLATE 200

632 | *Anthocleista grandiflora* [=*A. zambesiaca*] | Forest fever tree | Boskoorsboom
North-eastern Limpopo | January to June
Moist warm forest

A forest tree in the tropics with huge, simple leaves and a sparsely branched crown

PLATE 201

(90% of actual size)

486 | *Garcinia livingstonei* | African magosteen | Laeveldse geelmelkhout
North-eastern Limpopo, eastern Mpumalanga, north-eastern KwaZulu-Natal | November to December
Open woodland, riverine fringes and floodplains

An untidy tree with rigid branches rising at a sharp angle; leaves three-whorled

PLATE 202

(90% of actual size)

430.2 | *Pancovia golungensis* | False soap-berry | Basterseepbessie
North-eastern KwaZulu-Natal | December to February
Coastal and dune forest

Rare shrub or small tree; young leaves pinkish; flowers and fruit on old wood

688.5 | *Catunaregam taylorii* [=*C. spinosa* & =*Xeromphis obovata*] | Thorny bone-apple | Doringbeenappel
North-eastern Limpopo, Mpumalanga, north-eastern KwaZulu-Natal | April to May
Coastal bushveld, dense thicket, open woodland

A small tree; dark green leaves, densely hairy both sides; straight paired spines

PLATE 204

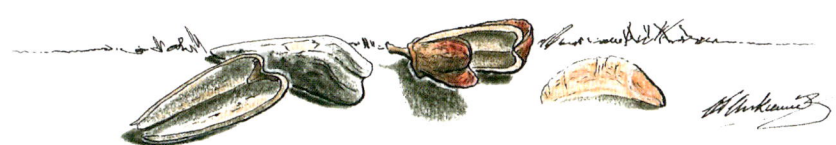

612 | *Schrebera alata* | Wing-leaved wooden-pear | Wildejasmyn
Limpopo, Mpumalanga, eastern KwaZulu-Natal | March to July
Coastal and montane evergreen forest, open woodland, bushveld

A small erect tree in rocky bushveld; leaves with a winged rachis is diagnostic.

PLATE 205

(85% of actual size)

682 | *Hymenodictyon parvifolium* | Yellow firebush | Geelbrandbos
Eastern Limpopo, eastern Mpumalanga | April to October
Open woodland, sandy soils, rocky ridges

A small tree or liana with bright yellow autumn colours; woody fruit persistent

PLATE 206

72 | *Brabejum stellatifolium* | Wild almond | Wildeamandel
Cape peninsula and mountains | February to May
Sheltered valleys, stream banks

A riverine tree, restricted to the south-west Cape with branches that can be enormous

Apricot size fruit

(Plates 207 to 227)

PLATE 207

506 | *Flacourtia indica* | Governor's plum | Goewerneurspruim
Limpopo, Gauteng, Mpumalanga | January to June
Woodland, riverine bush

A small tree with straight spines, particularly on coppice shoots and main trunk

PLATE 208

130.1 | *Capparis tomentosa* | Woolly caper-bush | Wollerige kapperbos
North-eastern Limpopo, Mpumalanga, eastern KwaZulu-Natal, north-eastern Eastern Cape | December to March
Evergreen and coastal forest, mountain slopes, open woodland

A small tree or climber; young parts hairy; hooked, stipular spines devilish

PLATE 209

(90% of actual size)

580 | *Chrysophyllum viridifolium* | Fluted milkwood | Bosstamvrug
Coastal KwaZulu-Natal, north-eastern Eastern Cape | February to March
Coastal and evergreen forest

A canopy tree with fluted trunk; young leaves bronze; fine herring–bone venation

PLATE 210

(90% of actual size)

488 | *Warburgia salutaris* | Pepper-bark tree | Peperbasboom
Northern Limpopo, Mpumalanga, north-eastern KwaZulu-Natal | October to January
Evergreen forest, wooded ravines

Rare tree because it has been decimated by bark-collectors. Leaves are very peppery.

PLATE 211

(75% of actual size)

360 | *Sclerocarya birrea* [=*S. caffra*] | Marula | Maroela
North West, Limpopo, Mpumalanga, KwaZulu-Natal | April to June
Open woodland and bushveld

Big bushveld tree with compound leaves; leaflets with a long, pointed drip-tip

PLATE 212

(75% of actual size)

50 | *Ficus sur* [=*F. capensis*] | Broom cluster-fig | Besemtrosvy
Eastern Limpopo, Mpumalanga, eastern KwaZulu-Natal, eastern Eastern Cape, Garden Route | October to March
Forests, open wooded grassland, stream banks, moist ravines

The only fig tree with leaves having scalloped margins; fruit mostly on "brooms"

PLATE 213

(75% of actual size)

216 | *Cordyla africana* | Wild mango | Wildemango
Eastern Limpopo, north-eastern Mpumalanga | November to December
Hot dry bushveld, riverine forest, swamp forest, sandy soils

Deciduous trees mostly in sand forest with softly hanging foliage

PLATE 214

(75% of actual size)

133 | *Maerua cafra* | Common bush cherry | Gewone Witbos
Eastern North West, Limpopo, Mpumalanga, KwaZulu-Natal, coastal Eastern Cape | October to December
Open woodland, forest margins, dune bush

A small tree; leaves three to five foliate, leathery with hair-like tip

221

PLATE 215

(85% of actual size)

590 | *Vitellariopsis marginata* | Natal bush milkwood | Natalbosmelkhout
Coastal KwaZulu-Natal, north-eastern Eastern Cape | January to March
Coastal forest, forested ravines, river courses

A tall tree; greyish bark rough with rectangular cracks; leaves clustered apically

PLATE 216

(85% of actual size)

591 | *Inhambanella henriquesii* | Milk pear | Melkpeer
Northern coastal KwaZulu-Natal | October to November
Coastal and low altitude evergreen forest

Big coastal forest tree; young leaves spectacularly coppery-red

PLATE 217

(85% of actual size)

107 | *Monodora junodii* | Green apple | Groenappel
Extreme north-eastern Limpopo | January to March
Dry rocky outcrops, bushveld and sand forest

A small sand forest tree; twigs purplish; leaves soft and drooping

PLATE 218

146 | *Parinari curatellifolia* | Mobola plum | Grysappel
North-eastern Limpopo, eastern Mpumalanga | October to January
Bushveld, deciduous woodland, sandy soils

Tree with a dense canopy; bark rough and grey with deep rectangular blocks

PLATE 219

(90% of actual size)

105 | *Annona senegalensis* | Wild custard apple | Wildesuikerappel
Central Limpopo, Mpumalanga, north-eastern KwaZulu-Natal | December to March
Bushveld, sandy soils, river courses, rocky outcrops, swamp forest

Small tree; roundish leaves with strong lateral veins folded along the mid-rib

694 | *Rothmannia fischeri* | Woodland gardenia | Bosveldwitklokke
Extreme northern Limpopo | April onwards throughout the year
Bushveld, open woodland, sand forest and among rocks

A small evergreen tree; hairy pockets in the axils of the veins below prominent

PLATE 221

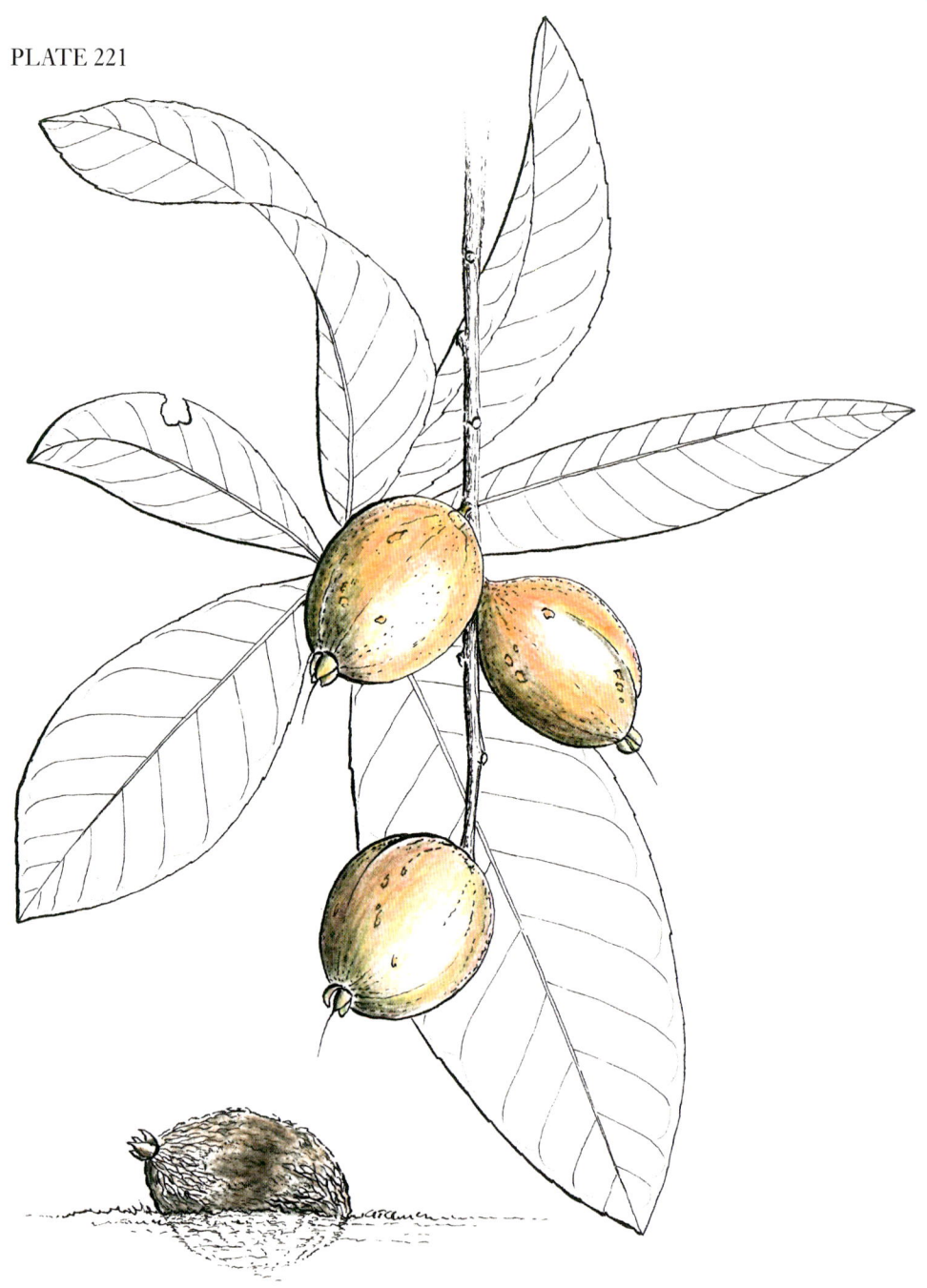

(85% of actual size)

524 | *Barringtonia racemosa* | Powder-puff tree | Poeierkwasboom
Coastal KwaZulu-Natal | July to October
River banks, fresh water swamps along estuaries

Trees of fresh or brakwater lagoons with large leaves forming dense stands

PLATE 222

(85% of actual size)

493 | *Xylotheca kraussiana* | African dog-rose | Afrikahondsroos
Far north-eastern Limpopo, coastal KwaZulu-Natal, north-eastern Eastern Cape | January to May
Coastal dune forest, sand forest and lowland bushveld

A small deciduous tree in forest–like situations with a pale grey, smooth bark

PLATE 223

(85% of actual size)

491 | *Rawsonia lucida* | Forest peach | Bosperske
North-eastern Limpopo, Mpumalanga, eastern KwaZulu-Natal, north-eastern Eastern Cape | November to February
Understorey of evergreen forest

A small tree with smooth mottled bark leaving orange patches

PLATE 224

(85% of actual size)

300 | *Trichilia dregeana* | Forest mahogany | Bosrooiessenhout
North-eastern Limpopo, Mpumalanga, eastern KwaZulu-Natal, north-eastern Eastern Cape | January to May
Coastal and montane forest

A big coastal forest tree with dark green shiny leaves

PLATE 225

(85% of actual size)

301 | *Trichilia emetica* | Natal mahogany | Rooiessenhout
Coastal KwaZulu-Natal | December to March
Riverine forest, woodland

A large handsome evergreen tree with a dense, spreading crown

PLATE 226

690.1 | *Gardenia cornuta* | Natal gardenia | Natalkatjiepiering
North-eastern KwaZulu-Natal | April to June
Open woodland, grassland, thicket

A multi-stemmed small tree in groves with smooth, yellowish-white mottled bark

PLATE 227

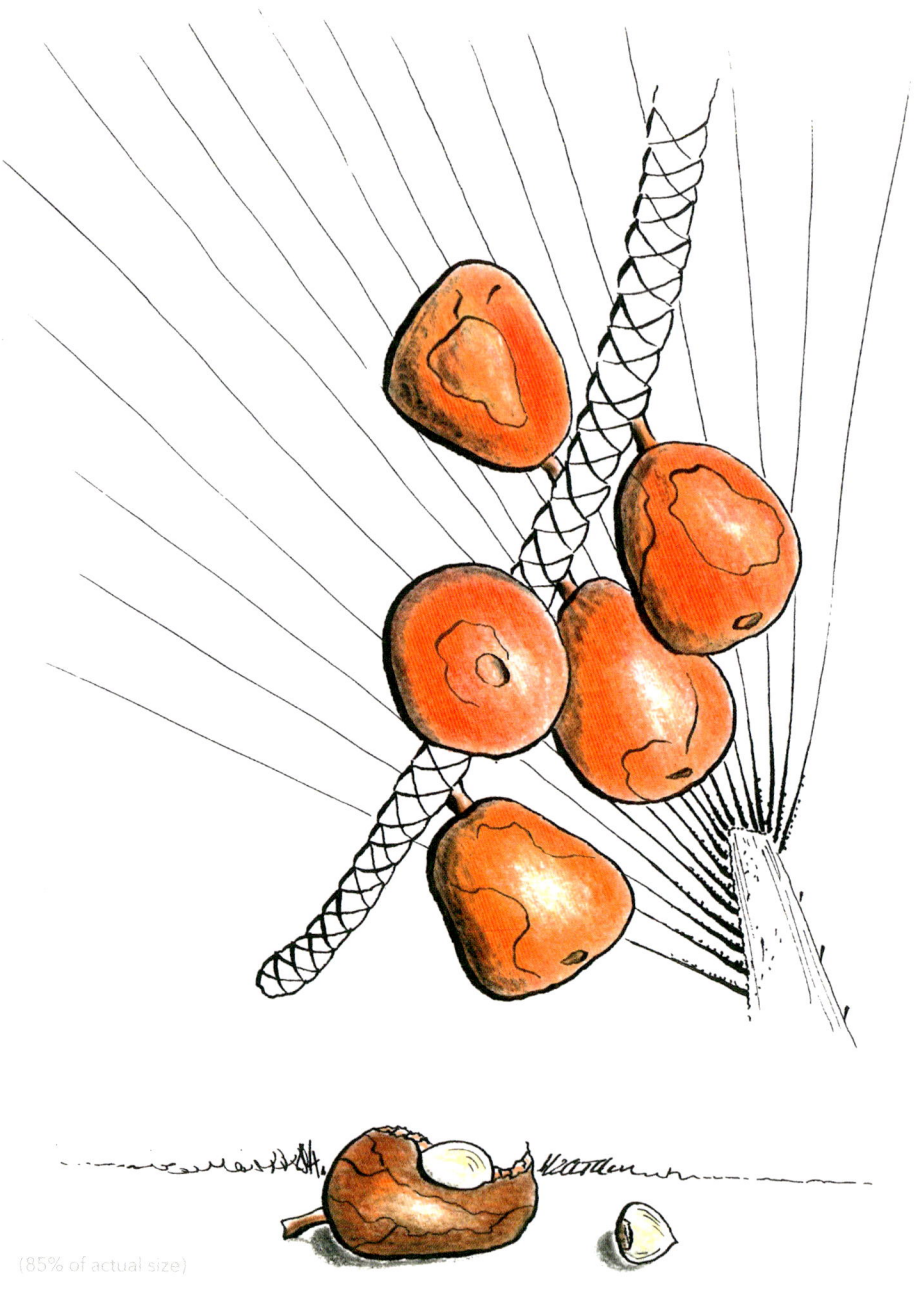

(85% of actual size)

23 | *Hyphaene coriacea* [=*H. natalensis*] | Ilala palm | Lalapalm
Eastern Mpumalanga, coastal KwaZulu-Natal | All year round
River banks, coastal woodland, sandy soils

Occurs from Satara southwards; often forming extensive stands on light soils

Peach size fruit

(Plates 228 to 230)

PLATE 228

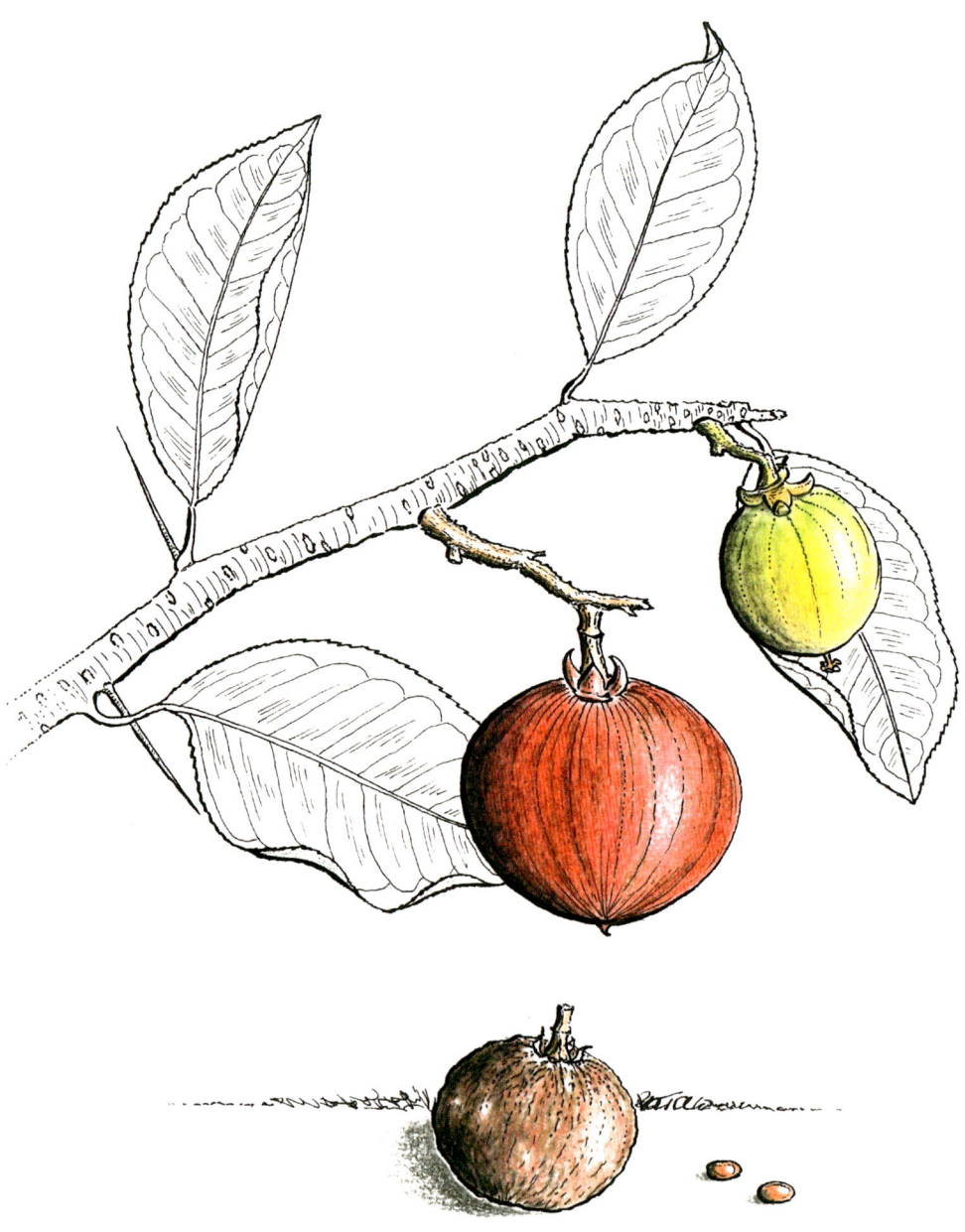

(85% of actual size)

492 | *Oncoba spinosa* | Snuff-box tree | Snuifkalbassie
Northern Limpopo, south-eastern Mpumalanga, north-eastern KwaZulu-Natal | April to July
Open woodland, riverine fringe forest, among rocks

A small tree with twigs covered in whitish lenticels; spines slender

PLATE 229

626 | *Strychnos madagascariensis* | Black monkey orange | Swartklapper
North West, Limpopo, Mpumalanga, coastal KwaZulu-Natal | February to November
Bushveld, rocky koppies, coastal forest

A small tree with thick, knobbly twigs; round leaves terminally clustered

PLATE 230

692 | *Gardenia thunbergia* | White gardenia | Witkatjiepiering
Coastal KwaZulu-Natal, coastal Eastern Cape | all year round
Evergreen forest, coastal thicket

A small tree in open forest, often multi-stemmed; smooth fruit persistent

Gem squash size fruit

(Plates 231 to 235)

PLATE 231

(75% of actual size)

693 | *Rothmannia capensis* | Cape gardenia | Kaapse katjiepiering
North West, Gauteng, north-eastern Limpopo, Mpumalanga, KwaZulu-Natal, Eastern and Western Cape | January onwards
Evergreen forest, wooded ravines, rocky hillsides

A forest tree in drier places; bark has a hessian-like texture

PLATE 232

(75% of actual size)

628 | *Strychnos pungens* | Spine-leaved monkey orange | Stekelblaarklapper
North West, south Limpopo, Gauteng | March to August
Bushveld, rocky places

A small tree with very tough, hard leaves that have a sharply pointed apex

PLATE 233

(85% of actual size)

623 | *Strychnos cocculoides* | Corky-bark monkey orange | Geelklapper
North West, western Limpopo, Gauteng | April to August
Bushveld, sandveld, rocky hillsides

A small bushveld tree with thick, corky bark; twigs end in a terminal spine.

PLATE 234

(75% of actual size)

691 | *Gardenia volkensii* [=G. spatulifolia] | Savanna gardenia | Bosveldkatjiepiering
North West, Limpopo, Gauteng, Mpumalanga, north-eastern KwaZulu-Natal | November to May
Bushveld, open woodland

A small bushveld tree; the persistent ribbed fruit are diagnostic.

PLATE 235

(50% of actual size)

467 | *Adansonia digitata* | Baobab | Kremetart
Northern Limpopo | April to May
Hot dry bushveld

Our largest, fattest tree occurring as a bushveld giant; fruit may be up to 200mm long

Pods straight

(Plates 236 to 267)

PLATE 236

(85% of actual size)

215 | *Peltophorum africanum* | African wattle | Huilboom
North West, Gauteng, Limpopo, Mpumalanga, north-eastern KwaZulu-Natal |
February to May
Wooded grassland, vlei margins, bushveld

Feathery foliage; yellow spring flowers; 'tree-like' stipules diagnostic

PLATE 237

(85% of actual size)

222 | *Bolusanthus speciosus* | Tree wisteria | Vanwykshout
Limpopo, north-eastern Mpumalanga, north-eastern KwaZulu-Natal |
February to March or later
Bushveld, wooded grassland, alkaline soils

A small graceful tree; Wisteria-like flowers; pendulous foliage; rough brown bark

PLATE 238

(85% of actual size)

232 | *Dalbergia melanoxylon* | Zebrawood | Sebrahout
North-eastern Limpopo, eastern Mpumalanga | January to March
Mixed woodland, thicket, rocky ridges

A small, stoutly-spiny deciduous tree with leaves clustered on short shoots

PLATE 239

(85% of actual size)

219 | *Calpurnia aurea* | Wild laburnum | Wildegeelkeur
Central Mpumalanga, KwaZulu-Natal, eastern Eastern Cape | Year round
Evergreen and riverine forest fringes, forest margins

A small tree; leaves drooping; hair-like tips; pale green and slivery-blue below

PLATE 240

(85% of actual size)

160 | *Senegalia ataxacantha* [=*Acacia ataxacantha*] | Flame thorn | Vlamdoring
North West, Gauteng, Limpopo, Mpumalanga, KwaZulu-Natal | June to October
Wooded grassland, open bush, sandy soils, rocky hillsides

A small tree or scrambler with very prickly stems; spectacular red pods

PLATE 241

226 | *Mundulea sericea* | Cork bush | Kurkbos
Gauteng, eastern Limpopo, Mpumalanga, north-eastern KwaZulu-Natal |
February to April
Bushveld, wooded grassland, rocky ridges, sandy flats

Small tree with corky bark and silvery leaves; often in small groves in bushveld

PLATE 242

221 | *Virgilia oroboides* | Blossom tree | Keurboom
Coastal Eastern and Western Cape | November to January or March to April
Forest margins, river valleys

A fast-growing pioneer after fire; often mistakenly identified as Black Wattle

PLATE 243

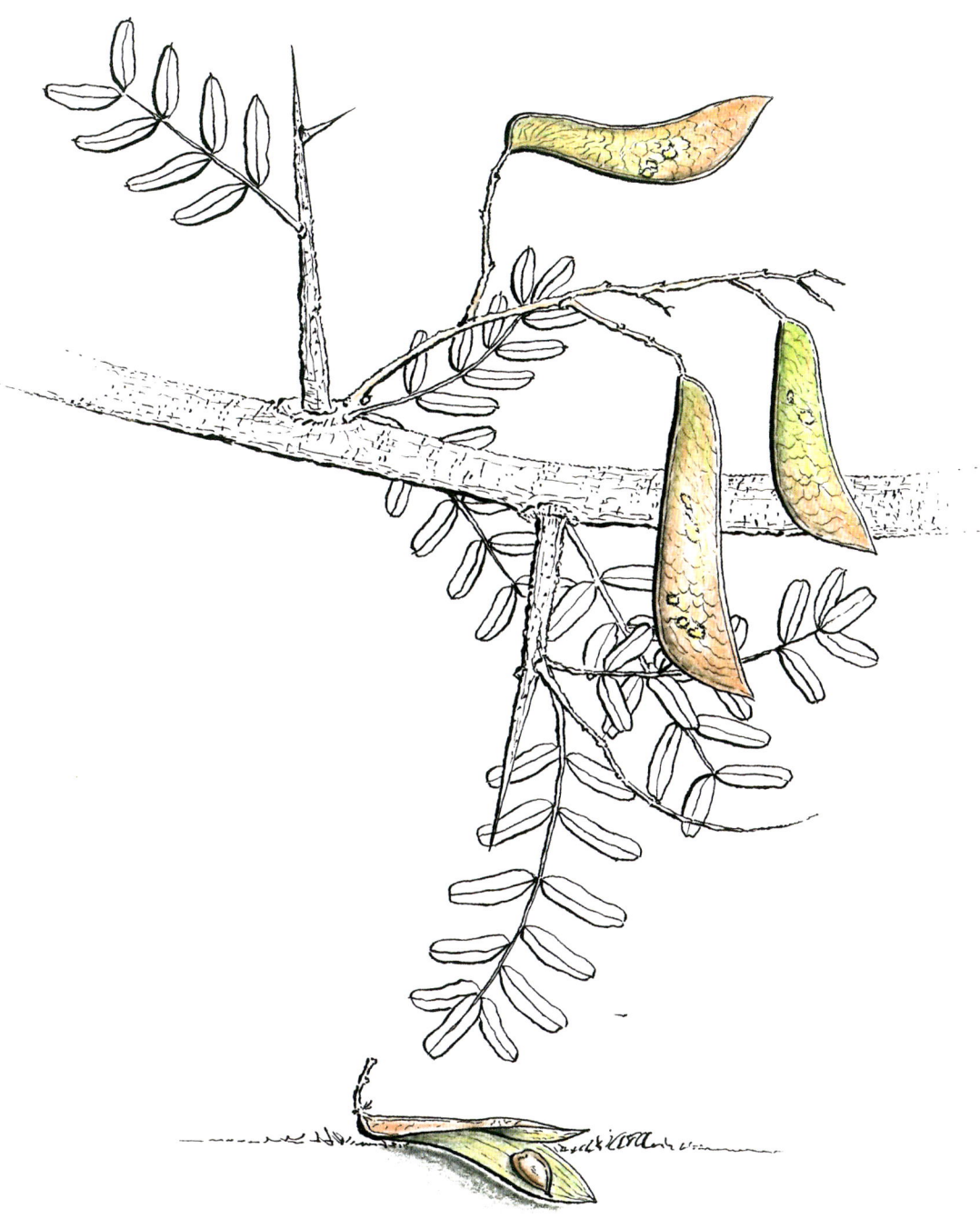

205 | *Umtiza listeriana* | Umtiza | Omtisa
East London (Buffalo Valley) | April to June
Deep wooded ravines

A spiny tree often mistaken for a Schotia *spp. when not in flower or fruit*

PLATE 244

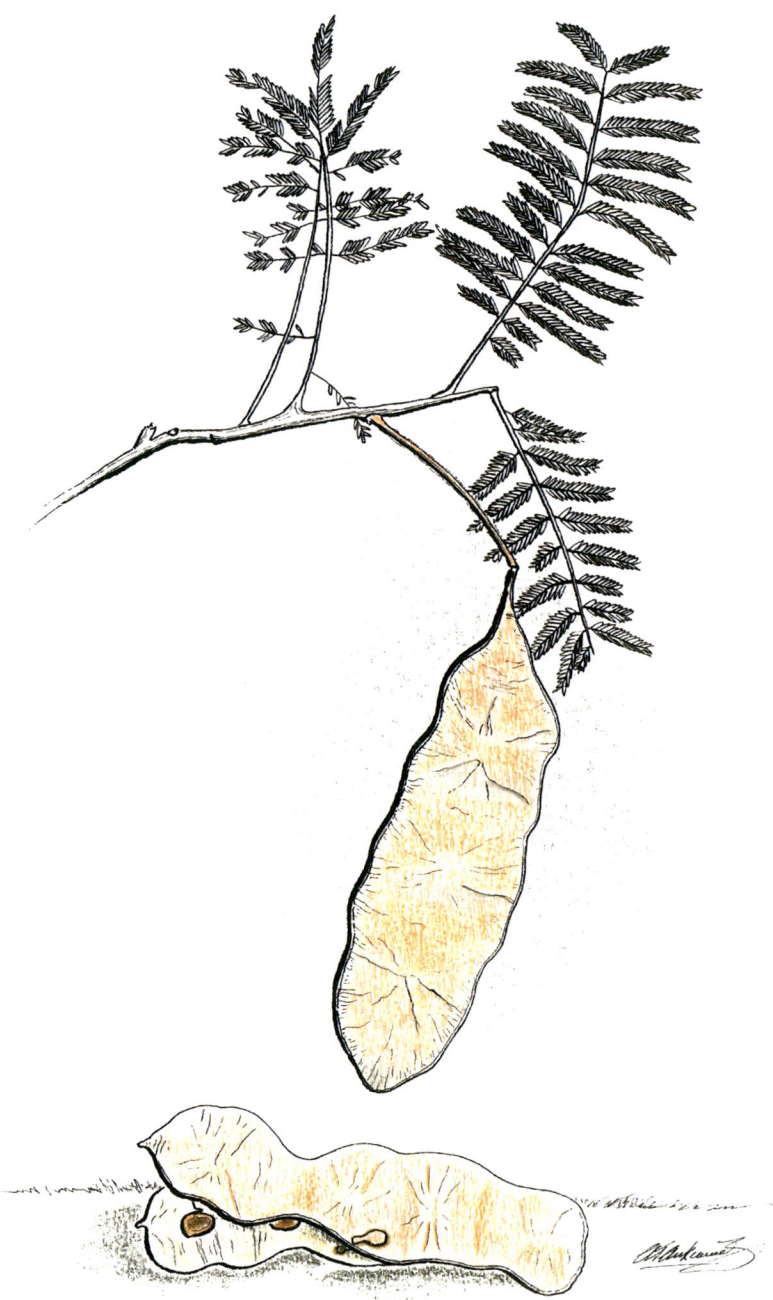

(85% of actual size)

154 | *Albizia forbesii* | Broad-pod albizia | Breëpeulvalsdoring
Eastern Mpumalanga, north-eastern KwaZulu-Natal | February to May
River banks, woodland, sand forest

A big tree with feathery foliage, a spreading crown and drooping branches

PLATE 245

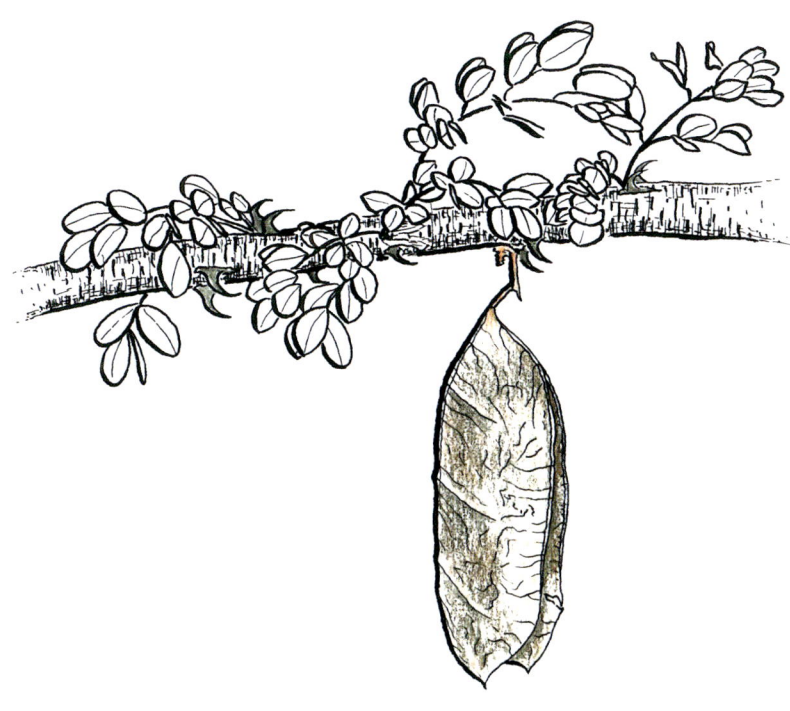

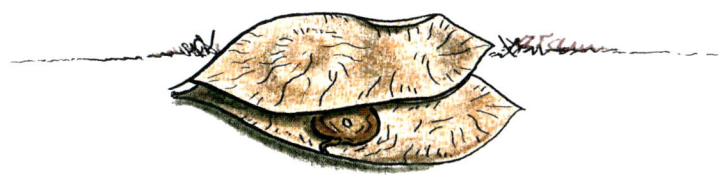

176 | *Senegalia mellifera* [=*Acacia mellifera*] | Black thorn | Swarthaak
Western Limpopo, North West, northern Northern Cape | January to April
Bushveld, semi-desert, Kalahari sands

A small tree; vicious paired, recurved thorns; leaflets quite large for an 'acacia'

174 | *Vachellia luederitzii* var. *luederitzii* [=*Acacia luederitzii* var. *luederitzii*] |
False umbrella thorn | Basterhaak-en-steek
Northern Northern Cape | February to May
Wooded grassland, thornveld, sandy soils

Small trees with a rounded canopy; hooked thorns; occurring in dense stands

PLATE 247

(75% of actual size)

197 | *Burkea africana* | Wild seringa | Wildesering
North West, Limpopo, Gauteng, western Mpumalanga | February to July
Bushveld and sandy soils

Large trees with flat, spreading crowns; twigs covered with reddish-brown hairs

PLATE 248

(70% of actual size)

238 | *Philenoptera violacea* [=*Lonchocarpus capassa*] | Apple-leaf | Appelblaar
Northern Limpopo, Mpumalanga | January to August
Woodland and river banks

Trees with sparse crowns and crooked, light brown trunks

PLATE 249

174.1 | *Vachellia luederitzii* var. *retinens* [=*Acacia luederitzii* var. *retinens*] |
Balloon thorn | Blaasdoring
Mpumalanga, north-eastern KwaZulu-Natal | March to June
Mountain slopes, dry river courses, stony flats

This subspecies is conspicuous because of the inflated thorns (see also Plate 246).

PLATE 250

170 | *Vachellia hebeclada* subsp. *hebeclada* [=*Acacia hebeclada* subsp. *hebeclada*] | Candle acacia | Trassiedoring
Limpopo, North West, North Cape | October to March
Hot, dry thicket and grassland

The unusual upright pods are diagnostic for this very thorny small tree.

PLATE 251

(75% of actual size)

202 | *Schotia brachypetala* | Weeping boer-bean | Huilboerboon
Limpopo, Mpumalanga, eastern KwaZulu-Natal, north-eastern Eastern Cape | February to May
Open deciduous woodland, scrub forest, river banks, bushveld

Big trees, often on termitaria, with slightly drooping, bright green foliage

PLATE 252

201 | *Schotia afra* var. *afra* | Karoo boer-bean | Karooboerboon
South-eastern Eastern Cape, Western Cape and far north Northern Cape |
October to March or later
Karroid scrub, rocky semi-desert, dry water courses

A small tree in the Karoo; the hard margins of the pods often persist as 'rings'

PLATE 253

(80% of actual size)

204 | *Schotia latifolia* | Bush boer-bean | Bosboerboon
Eastern Cape coastal area, Lebombo Mountains in Limpopo | April to August
Scrub forest, forest fringes, stony slopes

This small tree has pale pink or cream flowers and smooth reddish-brown bark.

(80% of actual size)

227 | *Millettia grandis* | Umzimbeet | Omsambeet
Coastal KwaZulu-Natal, coastal north-eastern Eastern Cape | June to September
Coastal forest and forest margins

Small trees; mature pods covered in rusty hairs that explode in hot weather

PLATE 255

(80% of actual size)

183 | *Vachellia robusta* subsp. *robusta* [=*Acacia robusta* subsp. *robusta*] |
River thorn | Enkeldoring
North West, western Limpopo, eastern North Cape, north-eastern Eastern Cape | October to February
Bushveld, wooded grassland

A tree with sturdy upward branches and a spreading crown

PLATE 256

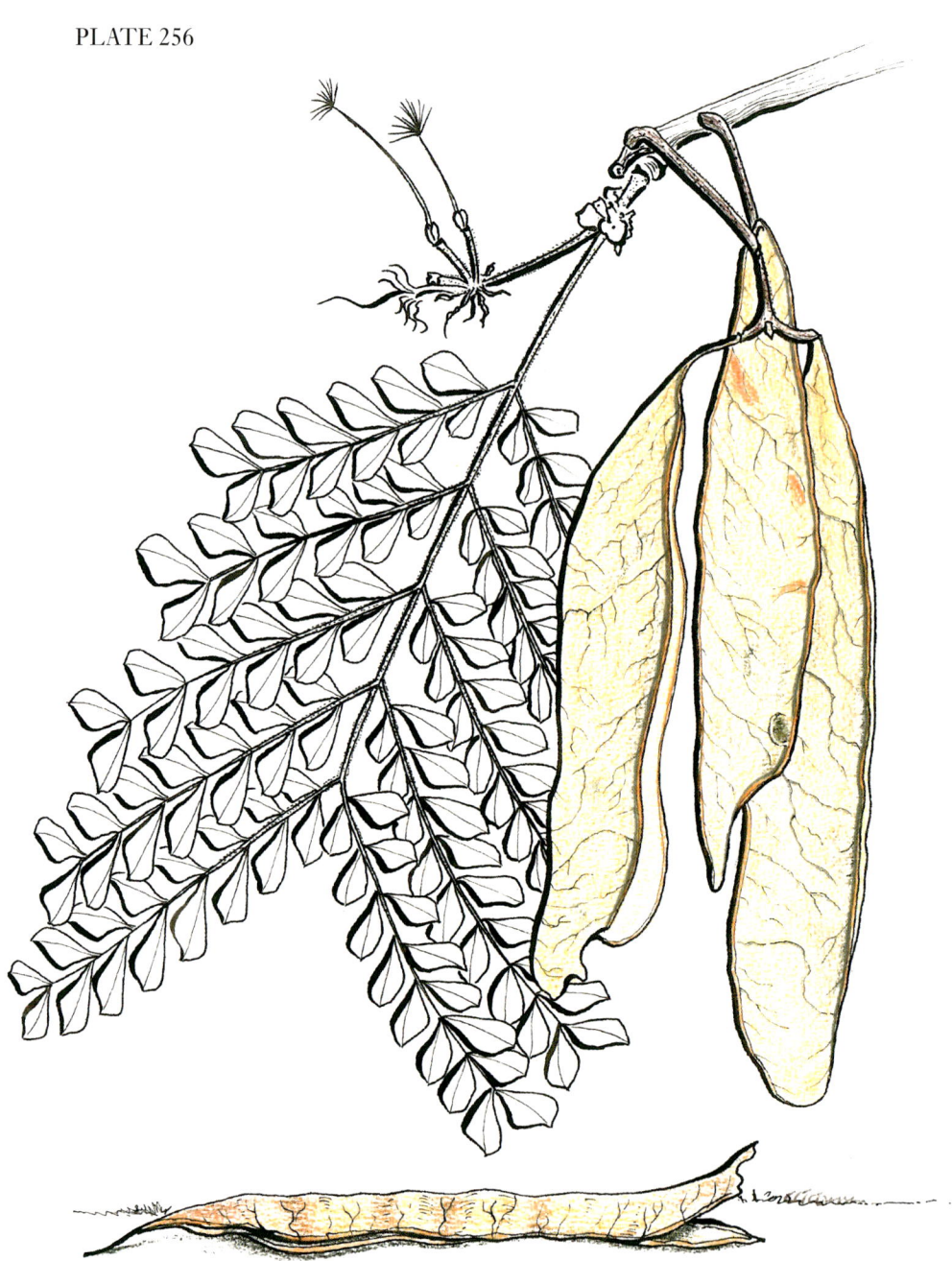

(90% of actual size)

148 | *Albizia adianthifolia* | Flat crown | Platkroon
North-eastern Limpopo, eastern Mpumalanga, eastern KwaZulu-Natal, north-eastern Eastern Cape | August to October
Woodland, coastal and montane forest, ravines

Large tree with a flat crown in forest-type situations; largish leaflets angular

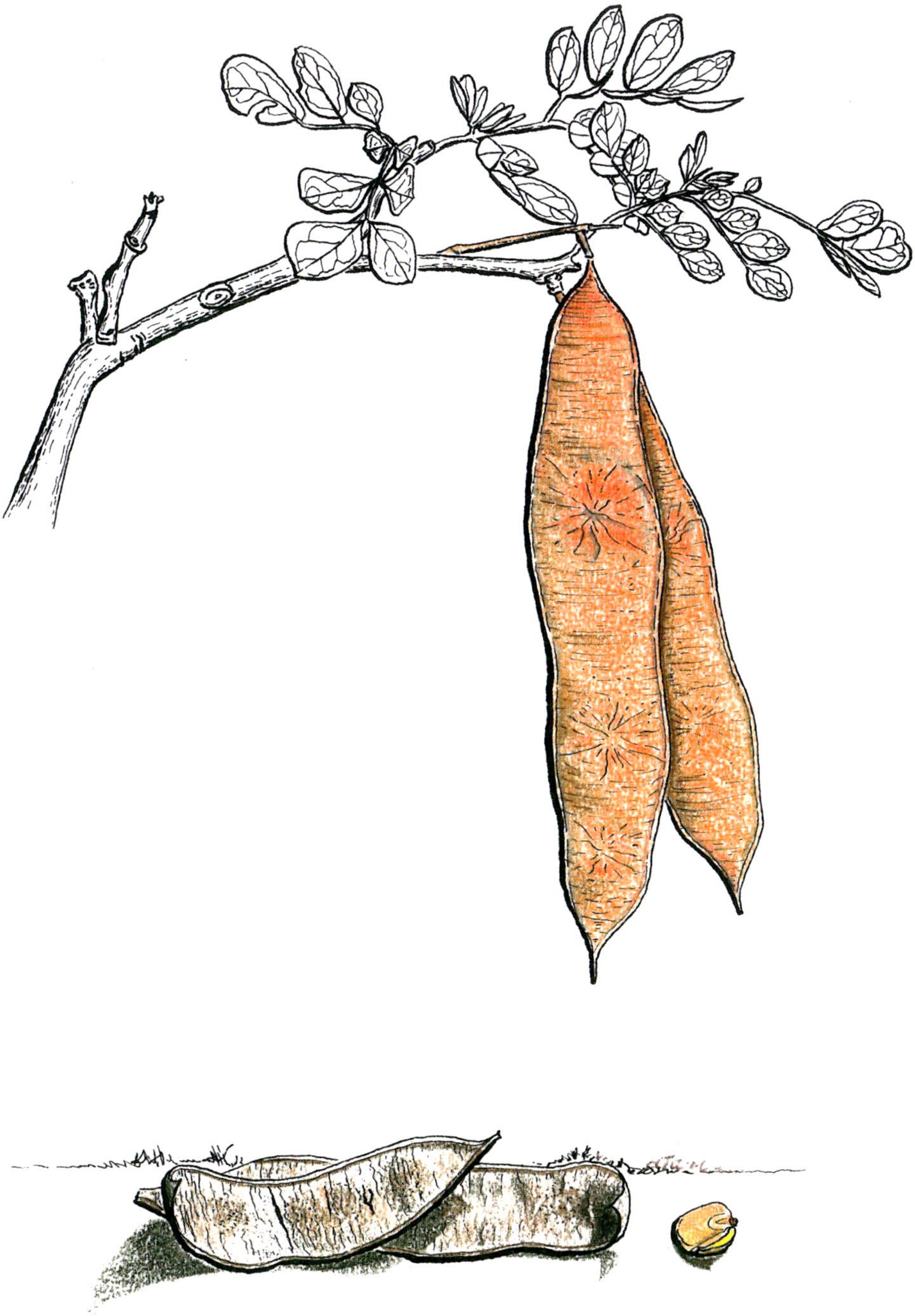

PLATE 257

153 | *Albizia petersiana* subsp. *evansii* | Many-stemmed albizia | Meerstamvalsdoring
Eastern Limpopo, eastern Mpumalanga | May
Open woodland, bushveld, sandy soils

Small, multi-stemmed upright trees occurring in groves in bushveld

PLATE 258

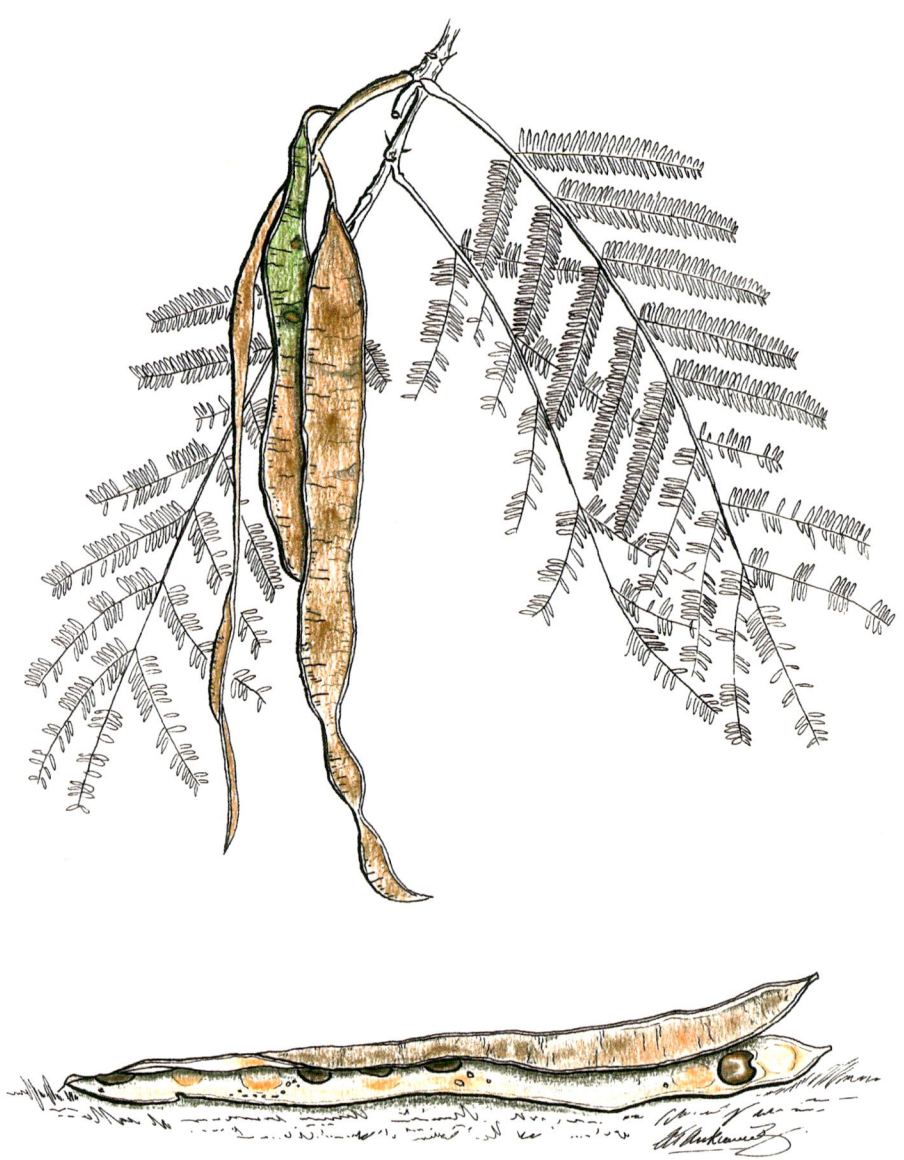

(80% of actual size)

162 | *Senegalia caffra* [=*Acacia caffra*] | Common hook-thorn | Gewone haakdoring
North West, Limpopo, Gauteng, Mpumalanga, KwaZulu-Natal, Eastern Cape coast | December to March
Woodland, wooded grassland, coastal scrub, stream banks

A small tree with fine, feathery foliage; our least thorny 'acacia'

PLATE 259

(80% of actual size)

178 | *Senegalia nigrescens* [=*Acacia nigrescens*] | Knob thorn |
Knoppiesdoring
Gauteng, Limpopo, eastern Mpumalanga, north-eastern KwaZulu-Natal |
January to June
Woodland, wooded grassland, river fringes

Leaflets the largest of our 'acacias'; spine-tipped knobs on the trunk are diagnostic.

170.2 | *Vachellia hebeclada* subsp. *tristis* [=*Acacia hebeclada* subsp. *tristis*] |
Candle thorn | Treurtrassiedoring
North Northern Cape | October to March
Dry grassland, bushveld thicket

Tree limited to the North West; unlike V. hebeclada *subsp.* hebeclada, *the pods are not upright.*

PLATE 261

(75% of actual size)

152 | *Albizia brevifolia* | Mountain albizia | Bergvalsdoring
Exreme north-eastern Limpopo | January to April
Dry stony hillsides, lowveld flats

A large tree or shrub; leaflets hairless when mature; A. harveyi *leaflets hairy*

PLATE 262

(75% of actual size)

161 | *Senegalia burkei* [=*Acacia burkei*] | Black monkey-thorn | Swartapiesdoring
North West, southern Limpopo, Mpumalanga, north-eastern KwaZulu-Natal | December to May
Bushveld, wooded grassland, sandveld

Lowveld bushveld tree with tough, sharply recurved, paired thorns

PLATE 263

(75% of actual size)

196 | *Erythrophleum lasianthum* | Swazi ordeal-tree | Swazi-oordeelboom
North-eastern KwaZulu-Natal | November to March
Dry bushveld, sand forest

Large trees with drooping foliage; leaves glossy; whole plant can be very toxic

PLATE 264

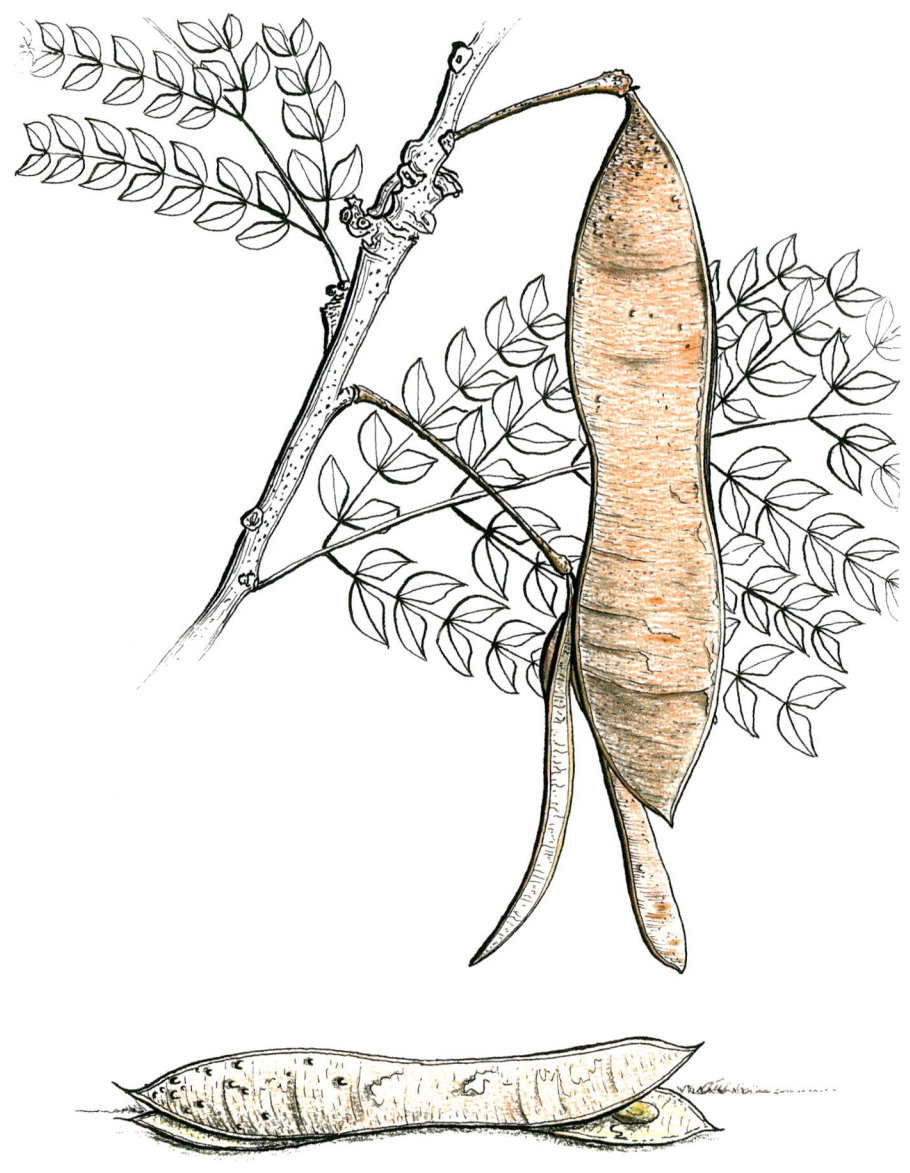

(70% of actual size)

Albizia gummifera | Smooth-bark albizia | Valsdoring
Restricted to eastern highlands of Zimbabwe | August to November
Montane forest

Large forest marginal trees resembling A. adianthifolia

PLATE 265

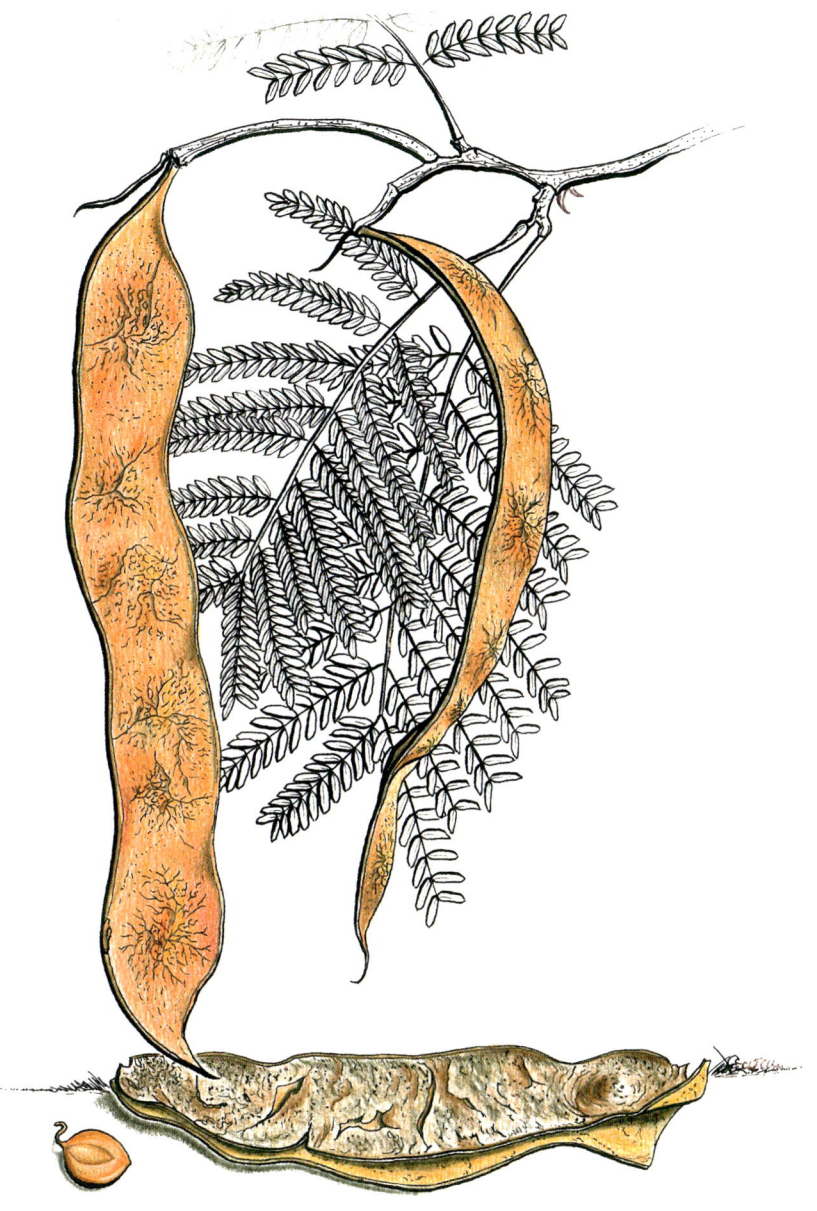

(70% of actual size)

166 | *Senegalia galpinii* [=*Acacia galpinii*] | Monkey thorn | Apiesdoring
Eastern North West, Limpopo | February to March
Wooded grassland, river banks

Medium to large tree; yellowish-brown bark and sharp, paired hooked thorns

PLATE 266

(75% of actual size)

187 | *Vachellia sieberiana* var. *woodii* [=*Acacia sieberiana* var. *woodii*] |
Paperbark thorn | Papierbasdoring
Limpopo, Mpumalanga, KwaZulu-Natal | March onwards
Woodland, wooded grassland, river courses, floodplains

Trees with a flattish canopy and heavy pods; white thorns paired and straight

PLATE 267

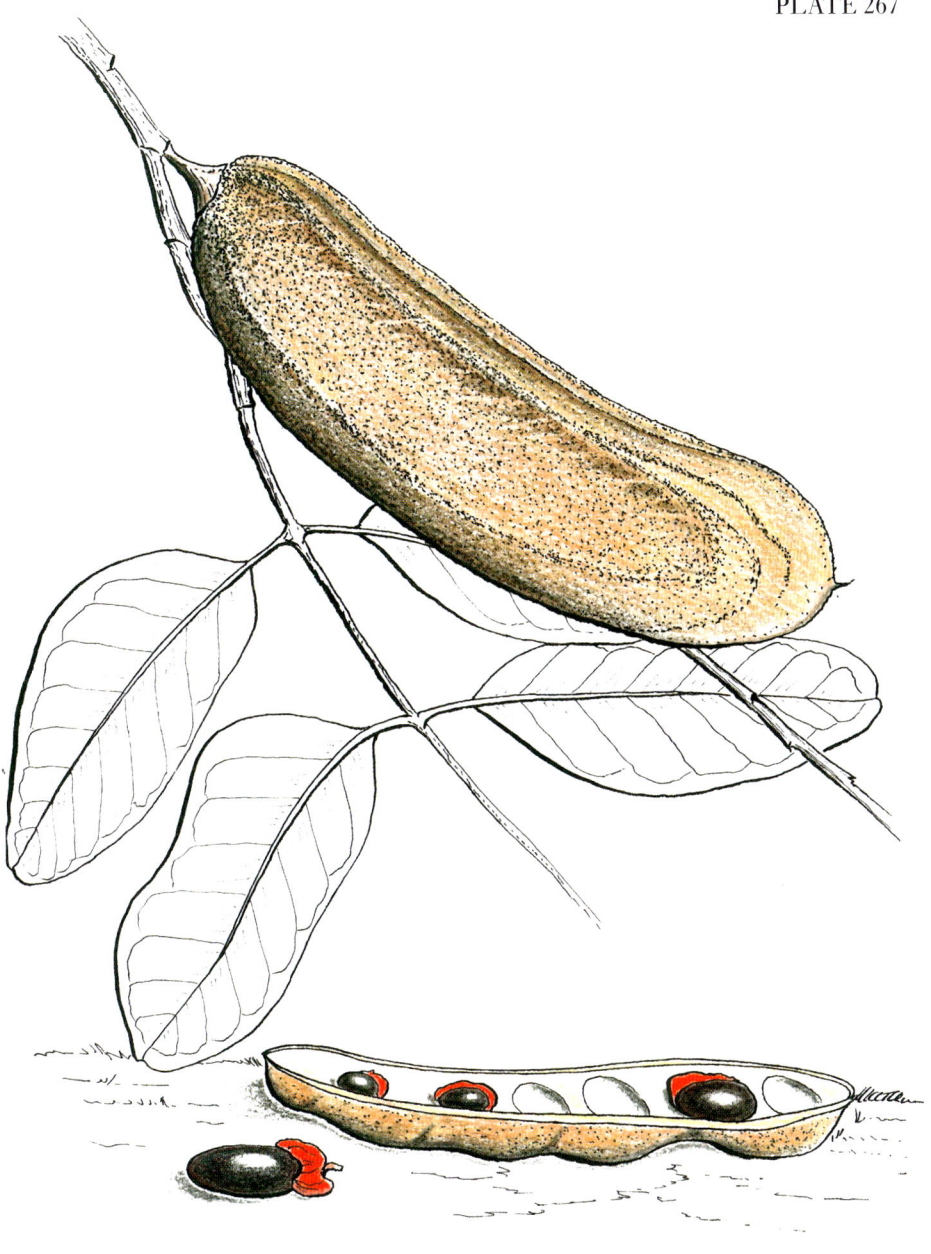

(75% of actual size)

207 | *Afzelia quanzensis* | Pod mahogany | Peulmahonie
Eastern Limpopo, Mpumalanga | June to November
Woodland and dry sand forest

Tall deciduous trees with thick pods and black seeds with a scarlet aril

Pods other

(Plates 268 to 285)

PLATE 268

(90% of actual size)

189 | *Vachellia xanthophloea* [=*Acacia xanthophloea*] | Fever tree | Koorsboom
Eastern Limpopo, eastern Mpumalanga, north-eastern KwaZulu-Natal |
January onwards
Moist swamp areas, river banks

Tall trees in groves with ascending branches and yellowish, smooth bark

PLATE 269

(90% of actual size)

214 | *Parkinsonia africana* | Wild green-hair tree | Wildegroenhaarboom
North-western Northern Cape | December to January
Desert, semi-desert, sandy soils near water courses

Small, spiny tree; leaflets fall early; the bare rachis makes the tree look 'hairy'.

PLATE 270

(80% of actual size)

109 | *Xylopia parviflora* | Tall bitterwood | Grootbitterhout
Extreme North Limpopo | All year
Evergreen forest, river fringe forest

A tall tree with a sparsely branched crown and narrow oval, drooping leaves.

PLATE 271

(80% of actual size)

179 | *Vachellia nilotica* subsp. *kraussiana* [=*Acacia nilotica* subsp. *kraussiana*] | Scented thorn | Lekkerruikpeul
North West, Gauteng, Limpopo, Mpumalanga, KwaZulu-Natal | March to September
Woodland, wooded grassland, scrub

Small, rounded tree with paired thorns that face slightly backwards

283

PLATE 272

(85% of actual size)

132 | *Maerua angolensis* | Bead-bean tree | Knoppiesboontjie
North West, northern Limpopo, north-eastern Mpumalanga, north-eastern KwaZulu-Natal | September to February
Wooded grassland, woodland, thicket

Leaves tend to be pendulous on long stalks with conspicuous bristle tips.

PLATE 273

(85% of actual size)

245 | *Erythrina lysistemon* | Common coral tree | Gewone koraalboom
North West, northern Limpopo, Mpumalanga, eastern KwaZulu-Natal, north-eastern Eastern Cape | December onwards
Scrub forest, dry woodland, coastal dune bush

A popular ornamental tree with prickly twigs and bright red to scarlet flowers

PLATE 274

(80% of actual size)

244 | *Erythrina latissima* | Broad-leaved coral tree | Breëblaarkoraalboom
Eastern Mpumalanga, KwaZulu-Natal, north-eastern Eastern Cape |
November to April
Wooded grassland, scrub forest, rocky slopes

A deciduous thickset tree with large leaflets and a thick, corky bark

PLATE 275

(80% of actual size)

167 | *Vachellia gerrardii* [=*Acacia gerrardii*] | Red thorn | Rooidoring
North Northern Cape, Limpopo, eastern Mpumalanga, KwaZulu-Natal |
December to May
Woodland, wooded grassland, river valleys

This flat-crowned tree has thick reddish twigs that are hairy when young.

PLATE 276

(85% of actual size)

172 | *Vachellia karroo* [=*Acacia karroo*] | Sweet thorn | Soetdoring
Widespread in all provinces | January onwards
Bushveld, grassland, coastal dune forest, river and stream banks

Our most common bushveld thorn tree with white, straight thorns

PLATE 277

(85% of actual size)

168.1 | *Vachellia grandicornuta* [=*Acacia grandicornuta*] | Horned thorn | Horingdoring
North West, Limpopo, Mpumalanga, north-eastern KwaZulu-Natal | April
Dry thornveld, Mopane woodland

Bushveld trees characterised by the 'boss' base to the paired thorns

PLATE 278

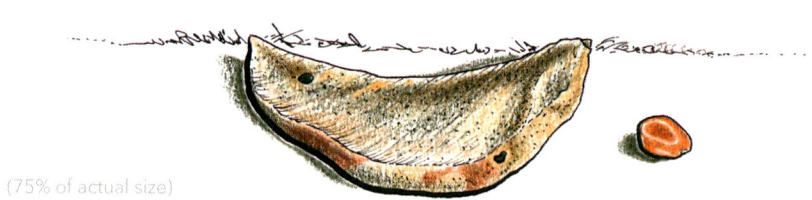

(75% of actual size)

168 | *Vachellia erioloba* [=*Acacia erioloba*] | Camel thorn | Kameeldoring
North West, Limpopo, northern Northern Cape | December to March
Dry woodland, grassland, stony sandy areas, water courses

Tree of the arid Kalahari with thick, ropy bark and large paired thorns

PLATE 279

(75% of actual size)

209 | *Piliostigma thonningii* | Camel's foot | Kameelspoor
North-eastern Limpopo, Mpumalanga | June to September
Bushveld, woodland, wooded grassland, sandveld

This small tree has large two-lobed leathery leaves resembling those of mopane.

PLATE 280

188 | *Vachellia tortilis* [=*Acacia tortilis*] | Umbrella thorn | Haak-en-steek
Nothern North West, Gauteng, Limpopo, Mpumalanga, eastern KwaZulu-Natal, North Cape | March to June
Woodland, grassland

A thorn tree with the flattest canopy and three kinds of paired thorns

PLATE 281

(75% of actual size)

190 | *Dichrostachys cinerea* | Sickle bush | Sekelbos
North Northern Cape, North West, Limpopo, Mpumalanga, eastern
KwaZulu-Natal | May to September
Bushveld, thicket

A small thorny tree, with spine-tipped branchlets easily confused with our 'acacias'

PLATE 282

(75% of actual size)

159 | *Faidherbia albida* [=*Acacia albida*] | Ana tree | Anaboom
Limpopo, Mpumalanga | September to October
Woodland, wooded grassland, river fringe forest, floodplains

A big tree that is summer deciduous; foliage tends to droop

PLATE 283

(75% of actual size)

169 | *Vachellia haematoxylon* [=*Acacia haematoxylon*] | Grey camel thorn | Vaalkameeldoring
North Northern Cape, western North West | January to April
Desert and semi-desert areas, deep red sandy soils, dry watercourses

This small Kalahari tree has the tiniest leaflets of all our 'acacias'.

PLATE 284

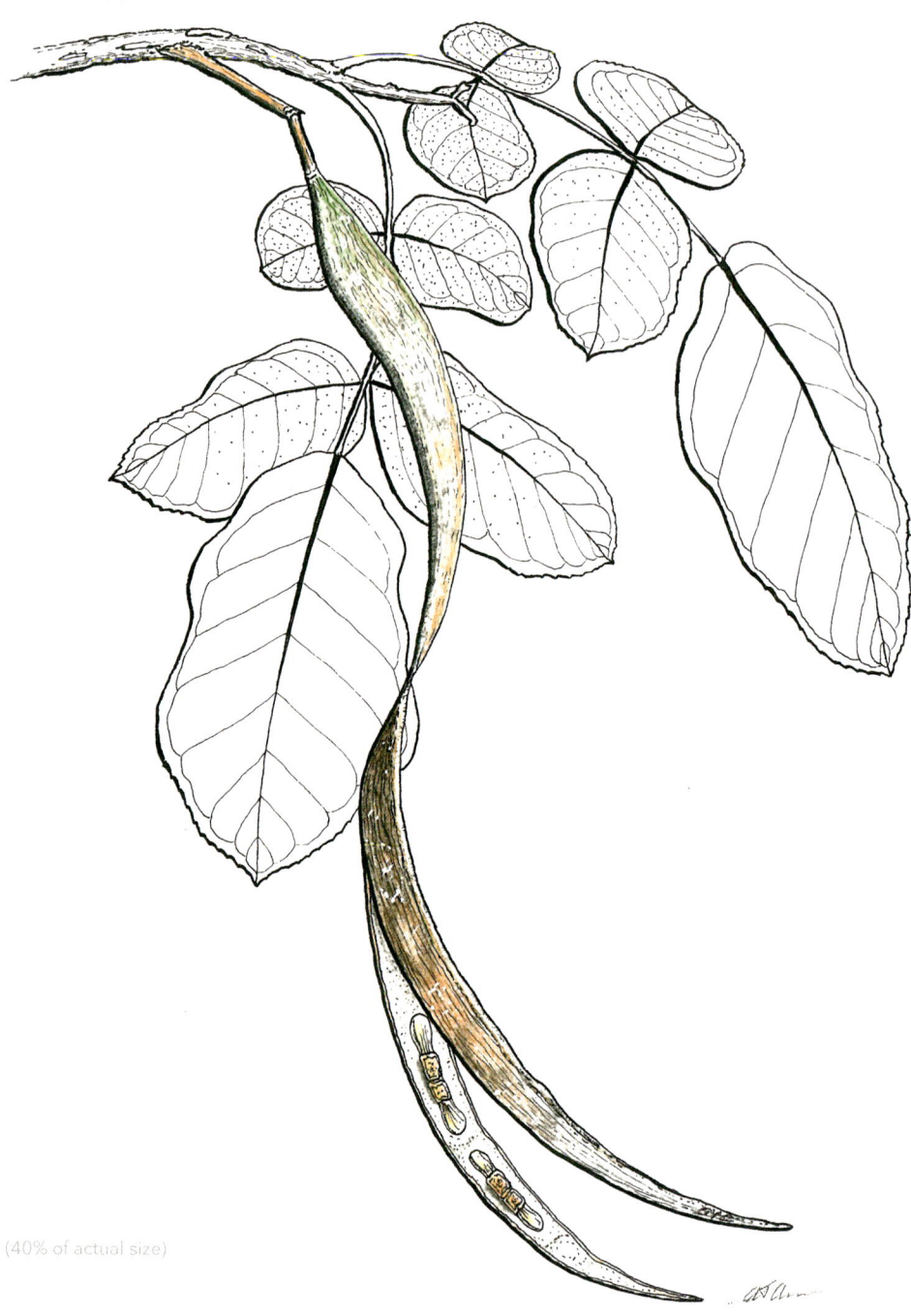

(40% of actual size)

677 | *Markhamia zanzibarica* [=M. acuminata] | Bean tree | Klokkiesboontjieboom
North-eastern Limpopo | January to May
Riverine fringes, rocky hillsides

Large pinnately compound, opposite leaves; narrow flaky bark gives a shaggy appearance

PLATE 285

198 | *Colophospermum mopane* | Mopane | Mopanie
Extreme northern Limpopo | March to June
Hot low-lying alluvial soils

A dominant species in almost pure stands; butterfly-like leaves are diagnostic.

Fruit: winged, lobed, cylindrical

(Plates 286 to 334)

PLATE 286

303 | *Securidaca longipedunculata* | Violet tree | Krinkhout
North West, Limpopo and Mpumalanga | April to August
Woodland and bushveld

A small rare tree with leaves clustered on short, spiky shoots

PLATE 287

701 | *Alberta magna* | Natal flame bush | Breekhout
KwaZulu-Natal and north-eastern Eastern Cape | February to August
Evergreen forest margins, wooded ravines, rocky outcrops

A rare small tree; the brilliant winged fruit are as spectacular as the flowers

PLATE 288

(85% of actual size)

120 | *Gyrocarpus americanus* | Propeller tree | Helikopterboom
Extreme north Limpopo | July to September
Hot arid bushveld, stony hillsides

Tree with a smooth whitish trunk; three-lobed leaves, dark green above, hairy below

PLATE 289

(80% of actual size)

427 | *Atalaya alata* | Lebombo wing nut | Lebombokransesseboom
Northern coastal KwaZulu-Natal | February to May
Coastal scrub forest and rocky hillsides

Rare small tree mostly on scarps; asymmetric leaflets often almost sickle-shaped

PLATE 290

(80% of actual size)

237 | *Pterocarpus rotundifolius* | Round-leaved kiaat | Dopperkiaat
North-eastern Limpopo, Mpumalanga | November to April
Woodland, wooded grassland

A medium-sized tree; leaflets roundish with parallel veins prominent

236 | *Pterocarpus angolensis* | Wild teak | Kiaat
Eastern Limpopo, Mpumalanga, northern KwaZulu-Natal | January to April
Bushveld, wooded grassland, rocky hillsides

Flat crown trees; rough bark exudes a blood-red sticky sap when injured

PLATE 292

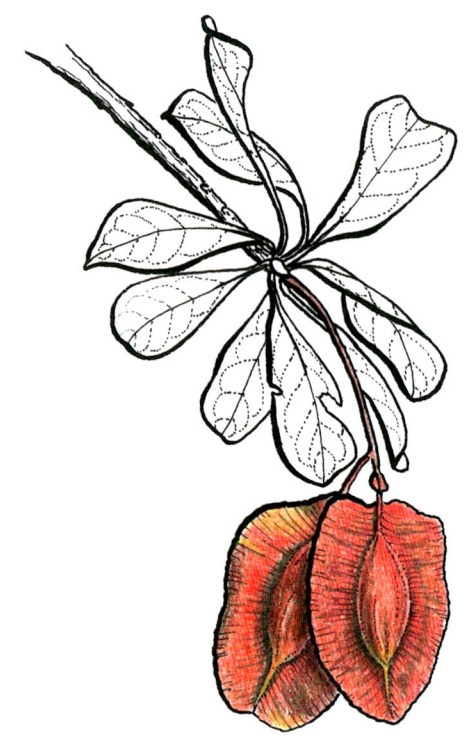

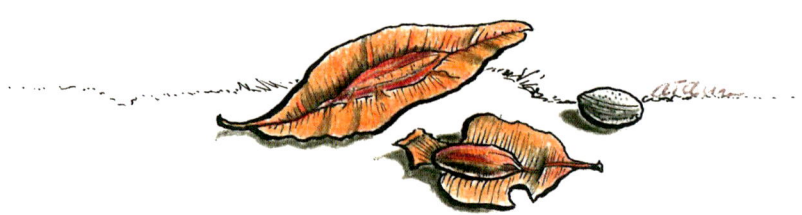

550 | *Terminalia prunioides* | Lowveld cluster-leaf | Sterkbos
Limpopo, Mpumalanga | January to July
Hot woodland, scrub, rocky hill slopes

A small, rather untidy tree, with outside branches drooping

PLATE 293

549 | *Terminalia phanerophlebia* | Lebombo cluster-leaf | Lebombotrosblaar
Eastern Mpumalanga, north-eastern KwaZulu-Natal | January to June
Bushveld, stony hillsides, water courses, forest margins

A small tree; leaves broadly obovate; prominent veins below; hairy when young

PLATE 294

551 | *Terminalia sericea* | Silver cluster-leaf | Vaalboom
North West, Limpopo, Mpumalanga, northern coastal KwaZulu-Natal |
January to May
Open woodland, sandy soils, vlei margins

Medium-sized tree with silvery-grey foliage; woody galls on twigs diagnostic

PLATE 295

437 | *Dodonaea angustifolia* [=*D. viscosa*] | Sand olive | Sandolien
Eastern KwaZulu-Natal, eastern Eastern Cape, Little Karoo, southern West coast | May to October
Riverine thicket, open woodland, rocky koppies

Erect small tree in groves; foliage shiny above; sticky towards the shoot tips

PLATE 296

540 | *Combretum kraussii* | Forest bushwillow | Bosvaderlandswilg
Limpopo, Mpumalanga, KwaZulu-Natal, north-eastern Eastern Cape |
February to June
Evergreen forest, wooded valleys, rocky places

A deciduous tree; conspicuous in spring as the whole canopy appears whitish

PLATE 297

537 | *Combretum molle* | Velvet bushwillow | Fluweelboswilg
North West, Limpopo, Gauteng, Mpumalanga, KwaZulu-Natal | January to June
Open woodland, sheltered rocky places

A small neat tree with dense velvety hairs on both leaf surfaces

PLATE 298

534 | *Combretum celastroides* subsp. *orientale* | Savanna bushwillow | Savanneboswilg
North-eastern Limpopo, Mpumalanga | February to June
Dry woodland, rocky hillsides, Kalahari sands

A small tree; wand-like twigs that twine; plum-coloured leaves in autumn

PLATE 299

536 | *Combretum erythrophyllum* | River bushwillow | Vaderlandswilg
North Northern Cape, western Free State, North West, Limpopo, Mpumalanga, KwaZulu-Natal, north-eastern Eastern Cape | January to October
River banks

Big trees with a spreading canopy and oft times in long, narrow groves

PLATE 300

534.1 | *Combretum edwardsii* | Natal combretum | Natalklimop
Central KwaZulu-Natal | January to July
Evergreen forest

A small tree or forest climber; young leaves turn red, gold and purple in autumn.

PLATE 301

538 | *Combretum hereroense* | Russet bushwillow | Kierieklapper
Northern Limpopo | February to June
Bushveld, open wooded grassland, vlei margins

A small neat tree with a dense crown and arching branches

PLATE 302

539 | *Combretum imberbe* | Leadwood | Hardekool
Northern North West, Limpopo, Mpumalanga | February to June or on to December
Bushveld, mixed woodland, along rivers and dry watercourses

Big trees with a greyish appearance; bark looks like 'crocodile-skin'.

PLATE 303

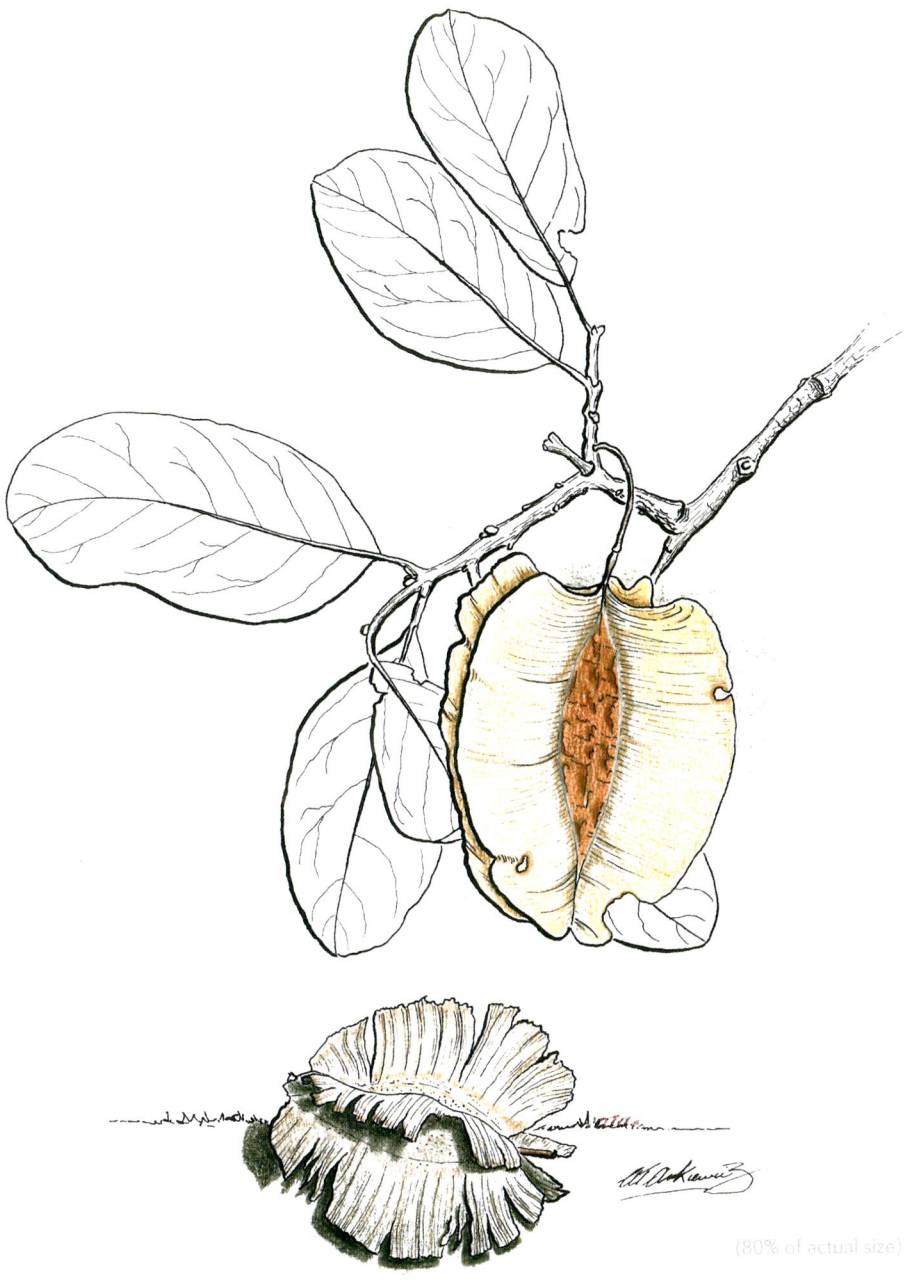

(80% of actual size)

546 | *Combretum zeyheri* | Large-fruited bushwillow | Raasblaar
North West, Gauteng, Limpopo, Mpumalanga, north-eastern KwaZulu-Natal | February to October
Bushveld, open woodland, rocky hillsides, river banks

Small to medium-sized tree with the largest of all four-winged fruit

PLATE 304

341 | *Spirostachys africana* | Tamboti | Tambotie
North West, Limpopo, Mpumalanga, eastern KwaZulu-Natal | October to December
Bushveld, river courses

Drainage line trees often in groves with blackish trunks and blocky bark

462 | *Grewia monticola* | Silver raisin | Vaalrosyntjie
North West, Limpopo, Mpumalanga, KwaZulu-Natal | February to August
Open woodland, riverine bush, sandy soils

Most attractive leaves of all the **Grewia** *spp.; bright green above, silvery white below*

PLATE 306

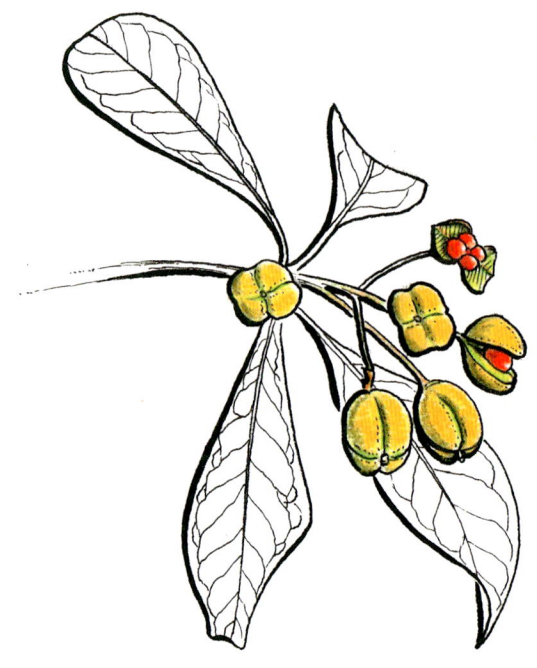

139 | *Pittosporum viridiflorum* | Cheesewood | Kasuur
Eastern Limpopo, Mpumalanga, KwaZulu-Natal, Eastern and West Cape |
May to September
Deciduous woodland, riverine thicket, rocky outcrops

Diagnostic is that if a leaf is removed, three dots appear on the separated leaf stalk.

PLATE 307

342 | *Shirakiopsis elliptica* [=*Sapium ellipticum*] | Jumping-seed tree | Springsaadboom
South-eastern Mpumalanga, coastal KwaZulu-Natal | March to August
Forest, wooded ravines, swamp forest

Medium-sized tree; leaf stalk joining the leaf blade from below is diagnostic.

PLATE 308

401.1 | *Maytenus procumbens* | Dune koko-tree | Duinekokoboom
Coastal KwaZulu-Natal, Eastern Cape coast, Garden Route | August to January
Coastal dune forest, woodland, wooded dune valleys, forest margins

A small tree; leaf margin thickened, rolled under and serrations spine-tipped

PLATE 309

711 | *Psydrax obovata* [=*Canthium obovatum*] | Quar | Kwar
Limpopo, Mpumalanga, eastern KwaZulu-Natal, eastern Eastern Cape,
Garden Route | March to May
Coastal forest, inland forest

A big unarmed tree; trunk twisted and fluted, bark blackish with longish flakes

PLATE 310

310 | *Margaritaria discoidea* | Pheasant-berry | Fisantebessie
Eastern Limpopo, eastern Mpumalanga, coastal KwaZulu-Natal, north-eastern Eastern Cape | December to February
Forest, bushveld, thicket, wooded grassland

Medium forest tree; a shrub in coastal bushveld; sometimes buttressed

PLATE 311

354 | *Euphorbia tetragona* | Honey euphorbia | Heuningnaboom
Eastern Cape coastal area | September to January
Dry thorn and scrub forest, valley bushveld

Succulent tree; stems mostly four-angled, but can be five; candelabra-like crowns

PLATE 312

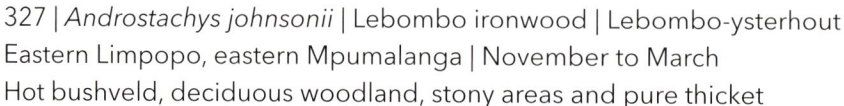

327 | *Androstachys johnsonii* | Lebombo ironwood | Lebombo-ysterhout
Eastern Limpopo, eastern Mpumalanga | November to March
Hot bushveld, deciduous woodland, stony areas and pure thicket

Medium-sized trees; leathery leaves dark green above, densely hairy below

PLATE 313

320 | *Cleistanthus schlechteri* | False tamboti | Bastertambotie
North-eastern KwaZulu-Natal | October to November
Open woodland, sand forest, riverine thicket

Medium-sized deciduous tree; drooping branches; reddish-brown autumn colours

PLATE 314

338 | *Suregada africana* | Common canary-berry | Gewone kanariebessie
South-eastern Mpumalanga, eastern KwaZulu-Natal, eastern Eastern Cape | November
Evergreen forest, bushveld

Small slender tree; the tiny raised dots on the shiny leaves are diagnostic.

PLATE 315

398 | *Maytenus acuminata* | Silky bark | Sybas
North-eastern Limpopo, Mpumalanga, KwaZulu-Natal, eastern Free State, coastal Eastern and Western Cape | May to October
Evergreen forest, forest margins, montane rocky outcrops, river banks

Small tree; diagnostic are silvery threads that hold a leaf together when snapped.

PLATE 316

307 | *Lachnostylis hirta* | Coalwood | Koolhout
Coastal Eastern and West Cape | February to May
Coastal bush, forest

A small tree; twigs with conspicuous leaf scars and leaf margins rolled under

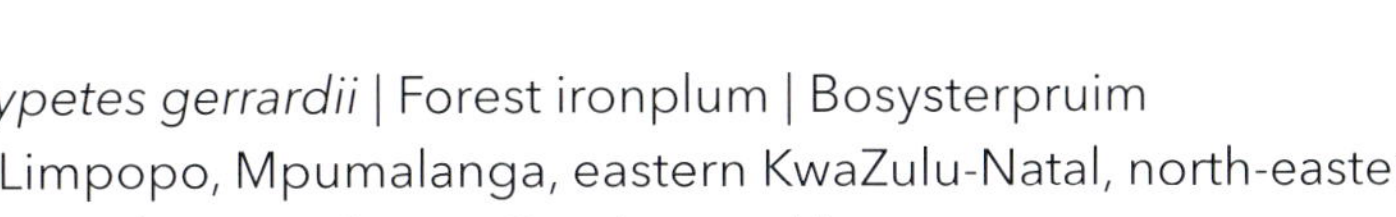

PLATE 317

314 | *Drypetes gerrardii* | Forest ironplum | Bosysterpruim
Eastern Limpopo, Mpumalanga, eastern KwaZulu-Natal, north-eastern Eastern Cape | September to October and later
Understorey tree in evergreen forest, open woodland, stream banks

Large tree; stem fluted and well buttressed; branches mostly horizontal

PLATE 318

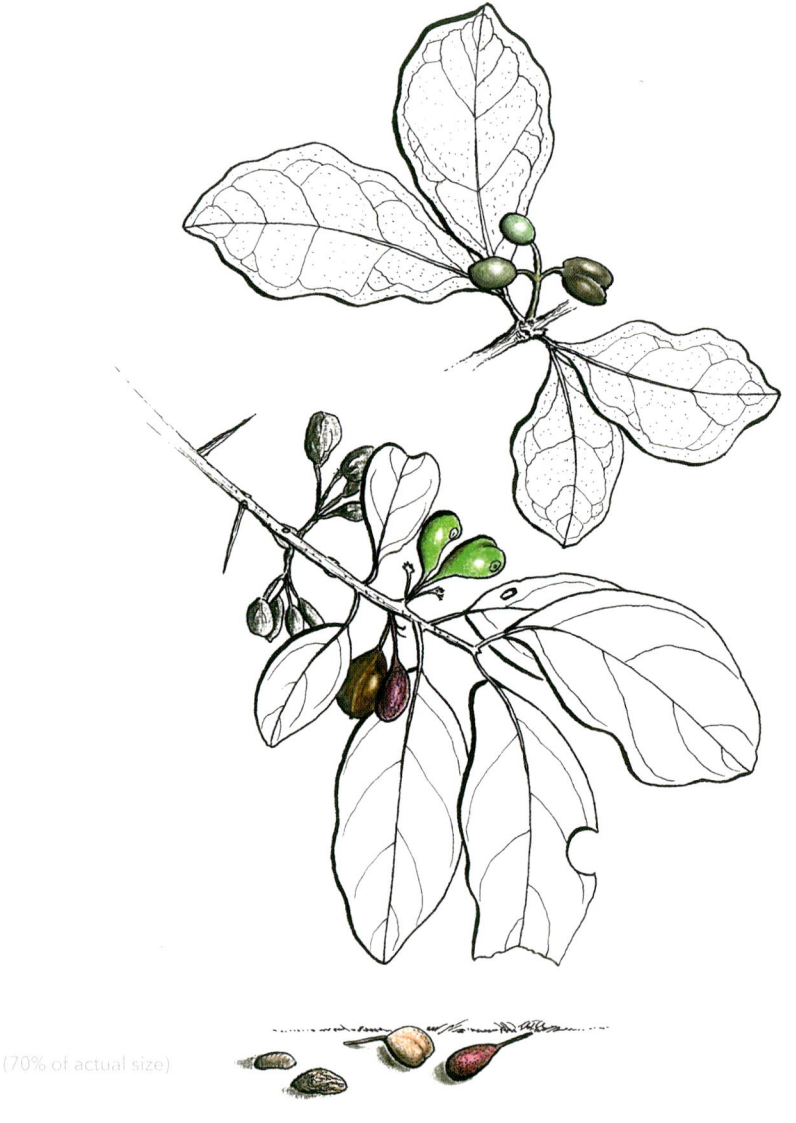

(70% of actual size)

706 | *Afrocanthium gilfillanii* [=*Canthium gilfillanii*] | Velvet rock-alder | Fluweelklipels (top); 708 | *Canthium inerme* [=*C. ventosum*] | Turkey berry | Gewone bokdrol (bottom)
North West, eastern Limpopo, Mpumalanga (top); Limpopo, Mpumalanga, KwaZulu-Natal, coastal Eastern and Western Cape (bottom) | November to May
Evergreen forest, forest margins, rocky outcrops; a pioneer in montane grassland

Turkey berry/Gewone bokdrol (bottom) very similar but thorny and absent in North West

PLATE 319

431 | *Smelophyllum capense* | Brittle bush | Buig-my-nie
Coastal Eastern Cape (Port Elizabeth, East London area) | December to January
Evergreen forest, ravines and bush

A rare small tree; the slender branches are very brittle.

PLATE 320

(90% of actual size)

306 | *Heywoodia lucens* | Stink ebony | Stinkebbehout
Coastal KwaZulu-Natal, north-eastern Eastern Cape | November onwards
Evergreen forest

Big trees in groves; dark foliage and dark bark; soft to the touch

PLATE 321

(80% of actual size)

30.9 | *Dracaena aletriformis* [=*D. hookeriana*] | Large-leaved dragon tree | Grootblaardrakeboom
Eastern Limpopo, eastern Mpumalanga, KwaZulu-Natal mountains to coast, Eastern Cape coast to Port Elizabeth | February to April
Dry shady places, dune forest, montane forest, sour bushveld

Trunks unbranched; strap-like leaves clustered at the end of stems

PLATE 322

(80% of actual size)

460 | *Grewia hexamita* | Giant raisin | Reuserosyntjie
Eastern Limpopo, Mpumalanga, north-eastern KwaZulu-Natal |
December to June
Deciduous woodland, river valleys

Small tree; bark smooth; lenticels conspicuous; branchlets with rusty-colour hairs

PLATE 323

(80% of actual size)

343 | *Sclerocroton integerrimus* [=*Sapium integerrimum*] | Duiker-berry | Duikerbessie
Eastern KwaZulu-Natal, north-eastern Eastern Cape | November to March and later
Open woodland, forest margins

Small tree, branches arched and weeping; seeds mottled (like castor oil bush)

PLATE 324

(90% of actual size)

329 | *Croton megalobotrys* | Large fever-berry | Grootkoorsbessie
North West, Limpopo, Mpumalanga | December to January
Alluvial flats, bushveld, riverine forest, thicket

Small-/medium-sized tree; twin stalked glands at the base of the leaf are diagnostic.

PLATE 325

225.4 | *Indigofera jucunda* | River indigo | Rivierverfbos
Coastal KwaZulu-Natal and north-eastern Eastern Cape | March to May
Dry hillsides, riverine forest, rocky grassland

A small neat tree; spectacular when in flower; leaflets dark green, hairless

PLATE 326

135 | *Maerua rosmarinoides* | Needle-leaved bush-cherry | Naadblaarwitbos
South-eastern Mpumalanga, eastern KwaZulu-Natal, north-eastern Eastern Cape | January to March
Dry thornveld and coastal scrubland

A small slender tree; very narrow trifoliolate leaflets are diagnostic.

641 | *Gonioma kamassi* | Kamassi | Kamassie
East Eastern Cape and Knysna forests | December to June
Evergreen forest near the coast

A small tree in forest understorey with four-whorled, shiny leaves

PLATE 328

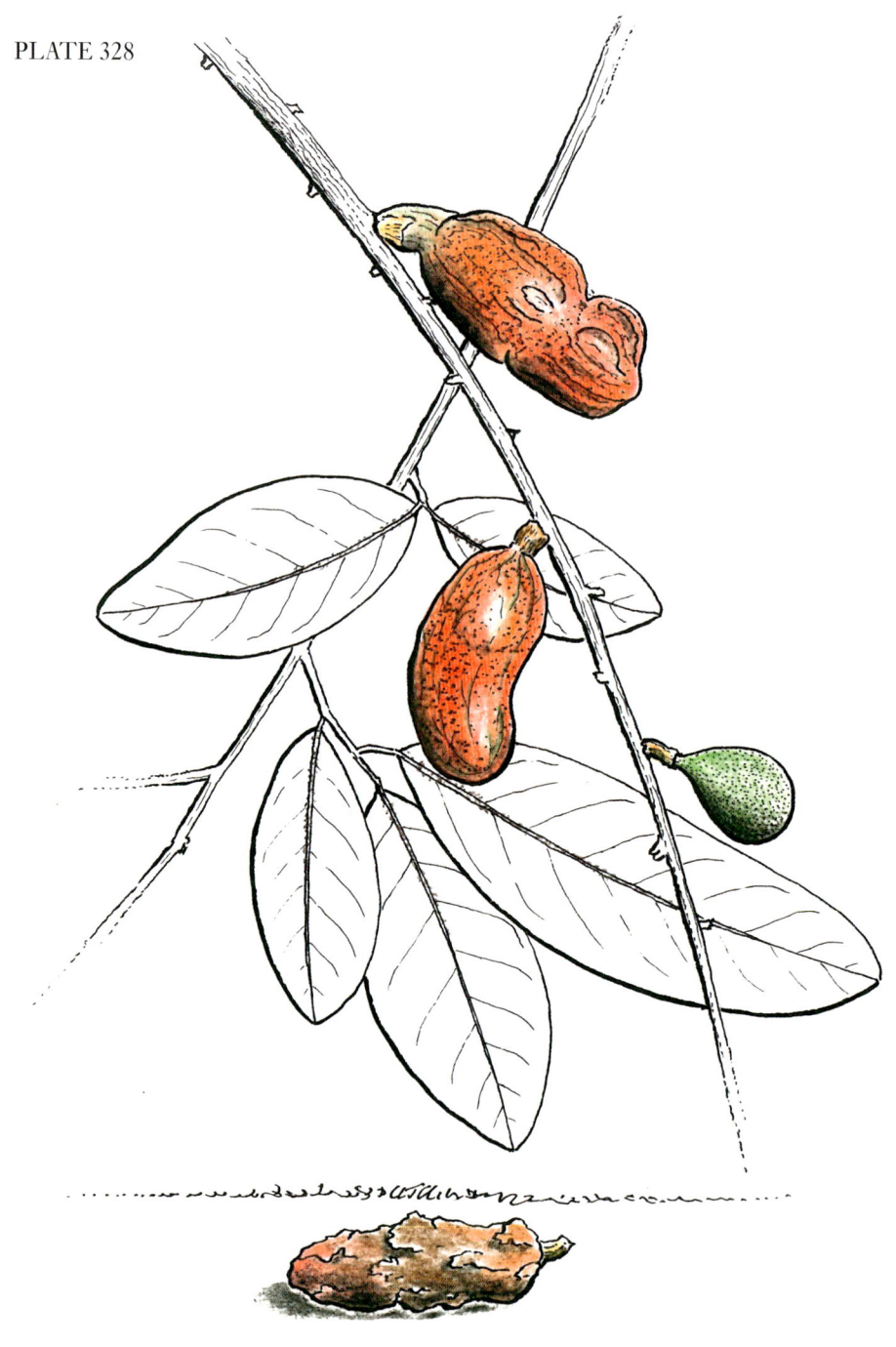

106 | *Hexalobus monopetalus* | Shakama plum | Shakamaprium
North-eastern Limpopo | December to April
Bushveld, rocky outcrops, riverine thicket

A small crooked tree; persistent petiole bases on old branchlets are diagnostic.

PLATE 329

527 | *Bruguiera gymnorrhiza* | Black mangrove | Swartwortelboom (main illustration); 526 | *Rhizophora mucronata* | Red mangrove | Rooiwortelboom (inset below left)
Both species in estuaries – Coastal KwaZulu-Natal, north-eastern Cape coast | September onwards
Seaward side of mangrove swamps

Black mangrove with cigar-shaped fruit and buttressed trunks; red with longer bulbous fruit and stilt roots

PLATE 330

(40% of actual size)

212 | *Cassia abbreviata* | Sjambok pod | Sambokpeul
North-eastern Limpopo, eastern Mpumalanga | December to April
Open woodland, wooded grassland, river courses, termite mounds

A medium-sized tree; pale green foliage; pods diagnostic and often persistent

PLATE 331

(40% of actual size)

650 | *Wrightia natalensis* | Saddle pod | Saalpeultjieboom
North-eastern Limpopo, eastern Mpumalanga, north-eastern KwaZulu-Natal | July to September
Arid woodland, sand forest, scrub hillsides

Medium-sized tree; spreading weeping crown; leaves exude a milky latex.

PLATE 332

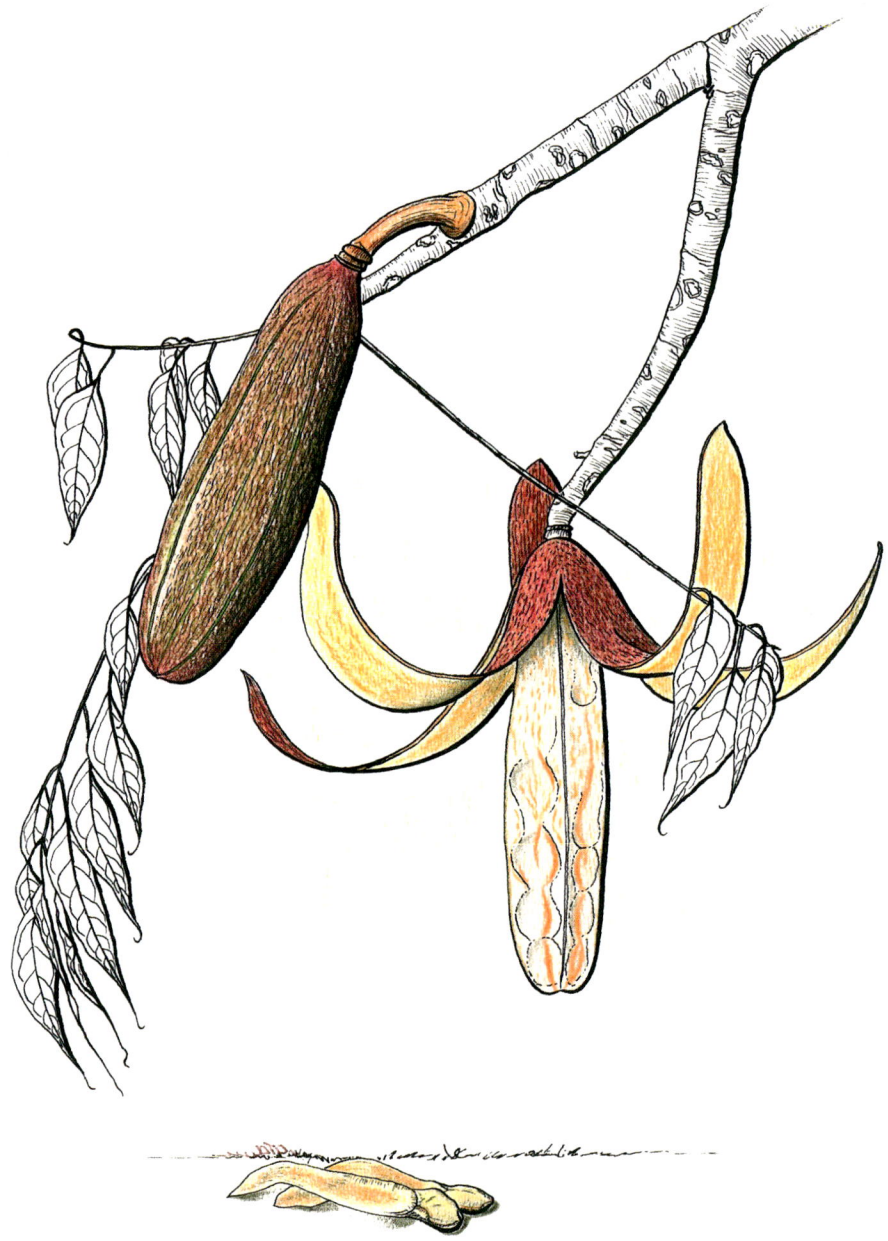

(75% of actual size)

293 | *Entandrophragma caudatum* | Wooden banana | Bergmahonie
Northern Limpopo, Mpumalanga | December to March
Dry bushveld, river valleys, rocky hillsides

A large tree; bark flakes in large scales; leaflets with long tapering tips

PLATE 333

(75% of actual size)

564 | *Cussonia spicata* | Cabbage tree | Gewone kiepersol
Widespread in eastern and southern regions of South Africa | June to September
Bushveld, forest margins, rocky outcrops in grassland

A thickset tree; blueish complex palmately compound leaves in an apical rosette

PLATE 334

(40% of actual size)

678 | *Kigelia africana* | Sausage tree | Worsboom
Eastern Limpopo, eastern Mpumalanga, eastern KwaZulu-Natal |
December to June
Open woodland, riverine bush

Large trees; sausages-like fruit can be 1-metre long and weigh several kilograms.

Other fruit

(Plates 335 to 381)

PLATE 335

(85% of actual size)

643 | *Diplorhynchus condylocarpon* | Horn-pod tree | Horingpeultjieboom
Northern North West, Limpopo | March to August
Bushveld, open woodland, rocky ridges, sandy soils

Small deciduous tree with copious milky latex; terminal twigs tend to droop.

PLATE 336

(85% of actual size)

478 | *Cola natalensis* | Common cola | Knuppelhout
Eastern KwaZulu-Natal, north-eastern Eastern Cape | February
Evergreen coastal forest

Medium-sized tree; leaves droop; leaf stalks with two swellings are diagnostic.

PLATE 337

(85% of actual size)

645 | *Tabernaemontana ventricosa* | Forest toad tree | Bospaddaboom
Eastern Mpumalanga, eastern KwaZulu-Natal | April to September
Evergreen coastal, montane forest

An understorey forest tree, the fruit being smooth not warty like T. elegans

PLATE 338

(70% of actual size)

644 | *Tabernaemontana elegans* | Toad tree | Paddaboom
Eastern Limpopo, Mpumalanga, north-eastern KwaZulu-Natal | April to September
Bushveld, coastal forest, river banks, rocky places

A small tree very similar to T. ventricosa *except fruit are warty*

PLATE 339

(80% of actual size)

477 | *Sterculia rogersii* | Common star-chestnut | Gewone sterkastaiing
Northern Limpopo and eastern Mpumalanga | September to March and later
Dry woodland, rocky outcrops

A small tree; trunk swollen; mottled bark shows distinct purplish patches

PLATE 340

(90% of actual size)

646 | *Voacanga thouarsii* | Wild frangipani | Wildefrangipani
Coastal KwaZulu-Natal | December to September
Evergreen, swamp forest, stream banks

Medium-sized tree; smooth bark; latex white; leaves crowded at branch ends

PLATE 341

404 | *Catha edulis* | Bushman's tea | Boesmanstee
Eastern Limpopo, Mpumalanga, western KwaZulu-Natal, central Eastern Cape | Mainly June to August
Bushveld, evergreen forest, open woodland, rocky wooded hillsides

Smallish tree resembling Eucalyptus *spp. from a distance; bark rough in older trees*

PLATE 342

269 | *Kirkia wilmsii* | Mountain syringa | Bergsering
North West, Limpopo, Mpumalanga | January to April
Dry bushveld, mountain slopes, rocky hillsides

Trees mainly on dolomitic soils; leaves crowded at the end of stout twigs

PLATE 343

267 | *Kirkia acuminata* | White seringa | Witsering
Far north-eastern Limpopo | January to next flowering season
Woodland, rocky ridges

Medium-sized deciduous tree; clean erect trunk; striking autumn colours

PLATE 344

(85% of actual size)

446 | *Greyia sutherlandii* | Natal bottlebrush | Natalse baakhout
Mpumalanga and inland KwaZulu-Natal | September to November
Grassy mountain slopes, rocky ridges

Small, gnarled tree; branches are brittle; showy red bottlebrush flowers

PLATE 345

295 | *Nymania capensis* | Chinese lanterns | Klapperbos
Little Karoo, central and western Northern Cape | October to December
Hot dry semi-desert, non-perennial water courses

Upright shrub or small tree; narrow linear leaves clustered on short shoots

PLATE 346

611 | *Diospyros whyteana* | Bladder-nut | Bostolbos
Widespread except in the dry arid west of South Africa | January to May
Mountain slopes, scrub forest, forest margins, rocky outcrops

A small understorey tree with mirror-like leaves fringed with ginger hairs

PLATE 347

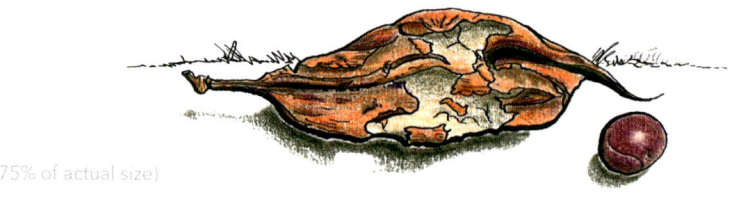

(75% of actual size)

436.2 | *Erythrophysa transvaalensis* | Bushveld red balloon | Bosveldrooiklapperbos
North West and western Limpopo | October to February
Bushveld and stony hillsides

A small deciduous tree; sparsely branched crown; leaf rachis slightly winged

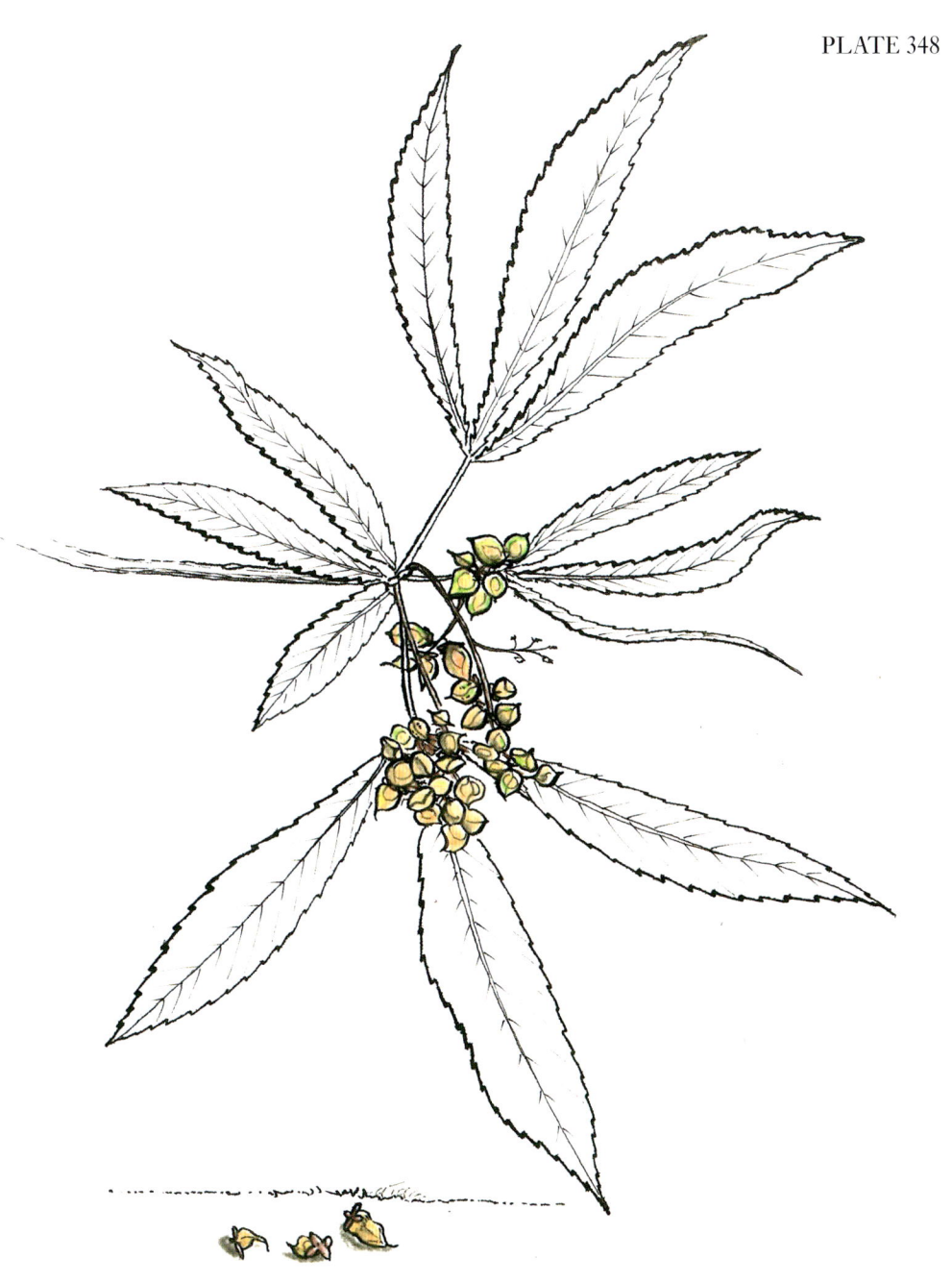

141 | *Platylophus trifoliatus* | White alder | Witels
Garden Route to the Cape Peninsula | April
Evergreen forests, stream banks

Medium-sized evergreen tree; clean bole or many-stemmed on old gnarled trunks

PLATE 349

292 | *Ptaeroxylon obliquum* | Sneezewood | Nieshout
North-eastern Limpopo, Mpumalanga, eastern KwaZulu-Natal, eastern Eastern Cape | December to February
Woodland, coastal forest, mist belt montane forest, sand forest

Medium-sized tree; leaves have a peppery smell when crushed.

PLATE 350

(80% of actual size)

569 | *Steganotaenia araliacea* | Carrot tree | Geelwortelboom
Eastern Limpopo, eastern Mpumalanga | November onwards
Hot dry woodland, rocky places

Small slender tree; white trunk; peeling bark; crushed leaves smell like carrot.

PLATE 351

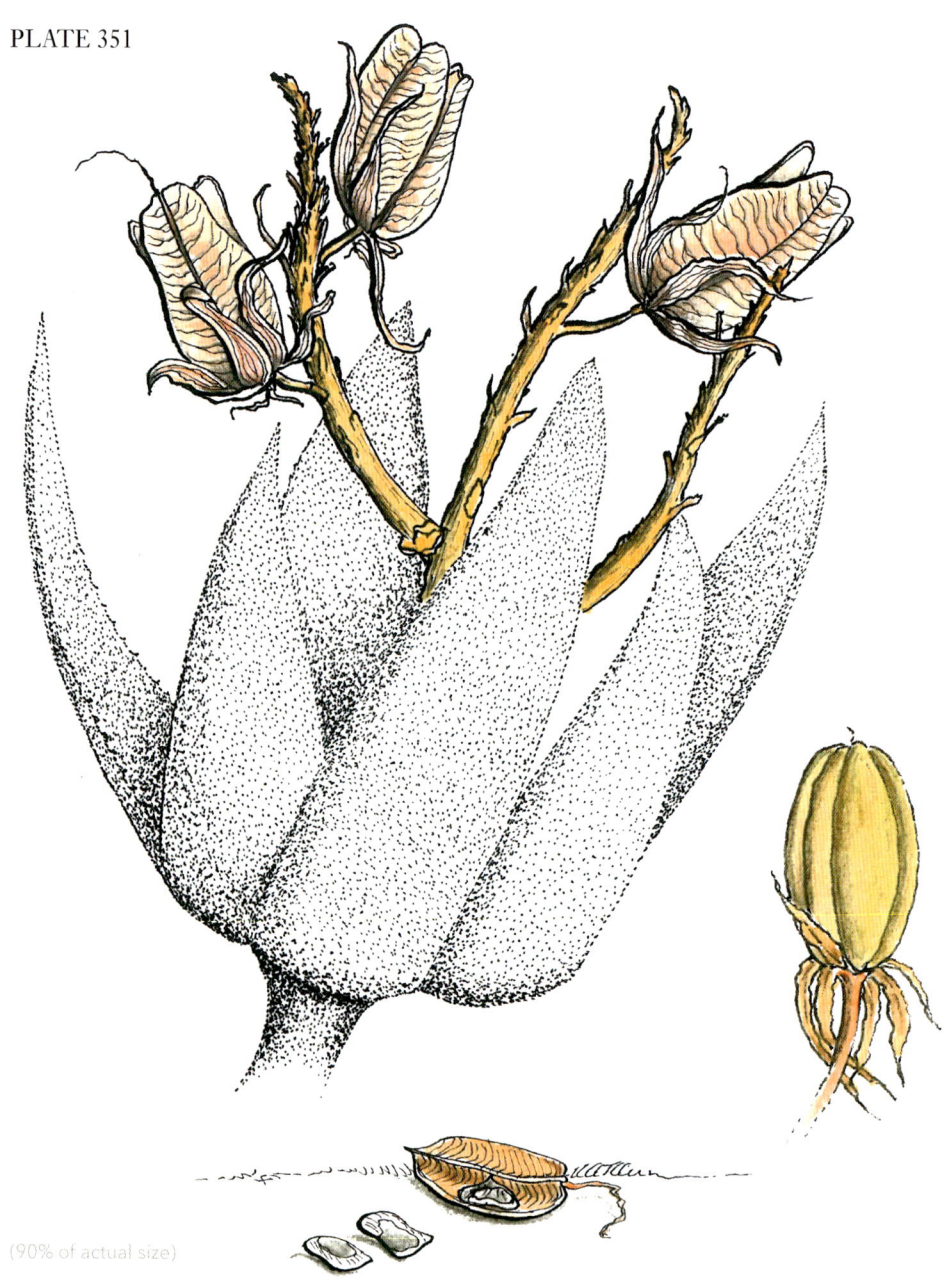

(90% of actual size)

29 | *Aloidendron dichotomum* [=*Aloe dichotoma*] | Quiver tree | Kokerboom; 28 | *Aloidendron barberai* [=*Aloe barberae, A. bainsii*] | Eastern tree-aloe | Boomaalwyn (inset)
North-western Northern Cape (main image); Eastern KwaZulu-Natal to east Eastern Cape (inset) | September to December (main image); July to September (inset)
Dry desert, semi-desert, rocky ridges (main image); Coastal bushveld (inset)

Main image: thickset tapering trunk; bark smooth but flaking with age. Inset: A. barberae *largest of the tree aloes to 20m*

(90% of actual size)

501 | *Homalium dentatum* | Brown ironwood | Bruinysterhout
Eastern Limpopo, Mpumalanga, eastern KwaZulu-Natal, north-eastern Eastern Cape | February to June
Evergreen forest, sheltered valleys, rocky hillsides

A medium-sized tree; smaller in scrub forest; trunk straight; drooping branches

PLATE 353

665 | *Vitex obovata* subsp. *wilmsii*. [=*V. wilmsii*] | Hairy fingerleaf | Harige vingerblaar
Mpumalanga, north-eastern KwaZulu-Natal | May to June
Evergreen forest, wooded hillside, coastal scrub, riverine bush

A small tree; three to five leaflets with short stalks; aromatic and can be hairy

PLATE 354

471 | *Dombeya rotundifolia* | Wild pear | Drolpeer
North-eastern North West, Limpopo, Mpumalanga, northern KwaZulu-Natal | October to December
Bushveld, wooded grassland

A small deciduous tree; rough leaves; spectacular when in flower

PLATE 355

521 | *Dais cotinifolia* | Pompon tree | Basboom
Eastern Limpopo, Mpumalanga, KwaZulu-Natal, north-eastern Eastern Cape | January to April
Evergreen forest margins, riverine forest, rocky mountain slopes

A small tree with a rounded crown; the smooth tough bark peels easily

PLATE 356

365 | *Loxostylis alata* | Tarwood | Teerhout
South-eastern KwaZulu-Natal, eastern coastal Eastern Cape | January to April
Forest margins, rocky outcrops, river banks, sandstone-derived soils

A small evergreen tree; leaves drooping and the rachis markedly winged

PLATE 357

608.1 | *Diospyros scabrida* | Coastal bladder-nut | Kusswartbas
Eastern KwaZulu-Natal, north-eastern Eastern Cape | March to April
Grassland, ravines, rocky outcrops

A small tree similar to D. whyteana *(Plate 346) except bracts open to reveal fruit.*

PLATE 358

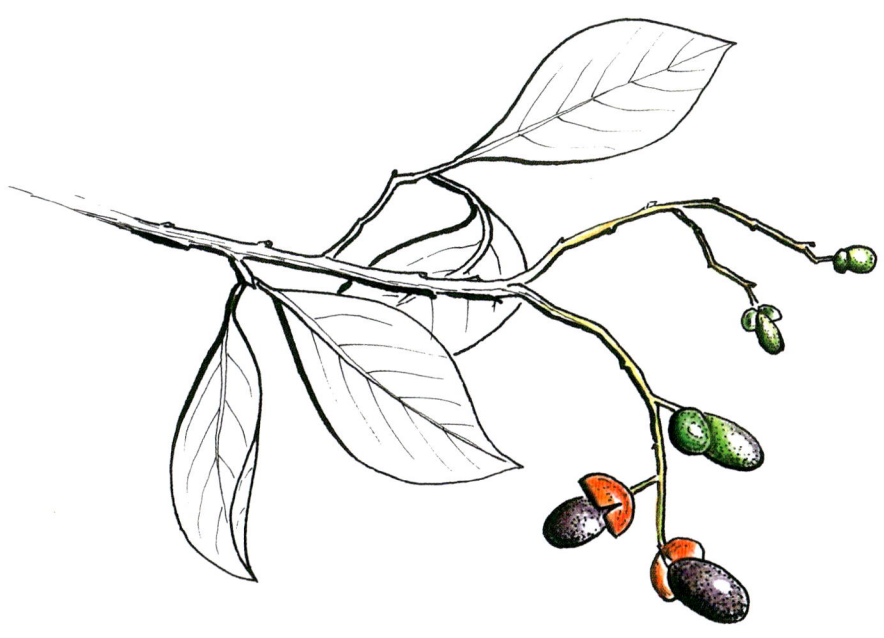

422 | *Apodytes dimidiata* | White pear | Witpeer
Eastern Limpopo, Mpumalanga, KwaZulu-Natal, Eastern and West Cape |
December to June
Coastal and inland forest; moist bushveld, rocky mountain slopes

A small to large tree; smooth grey bark; purplish young twigs; pinkish petioles

PLATE 359

479 | *Ochna arborea* | Cape plane | Kaapse rooihout
Eastern Limpopo, Mpumalanga, eastern KwaZulu-Natal, Eastern Cape to Garden Route | November to February
Evergreen forest, forest margins, stream banks

A slender small-/medium-sized tree; smooth, mottled bark; cool to the touch

PLATE 360

483 | *Ochna pulchra* | Peeling plane | Lekkerbreek
North West, Limpopo, western Mpumalanga | October to January
Open woodland, sandy soils, granite outcrops

Small trees often in groves; trunks smooth; peeling bark exposes creamy patches.

(85% of actual size)

15 | *Podocarpus elongatus* | Breede river yellowwood | Breëriviergeelhout (top); 18 | *P. latifolius* | Real yellowwood | Opregte geelhout (bottom)
P. elongatus Breede river endemic; P. latifolius eastern South Africa from Limpopo to Cape Town | August to February
P. elongatus Rocky stream banks; sandy soils. P. latifolius; evergreen coastal and montane forests

P. elongatus *a small bushy, resprouting tree.* P. latifolius; *a big upright forest tree*

PLATE 362

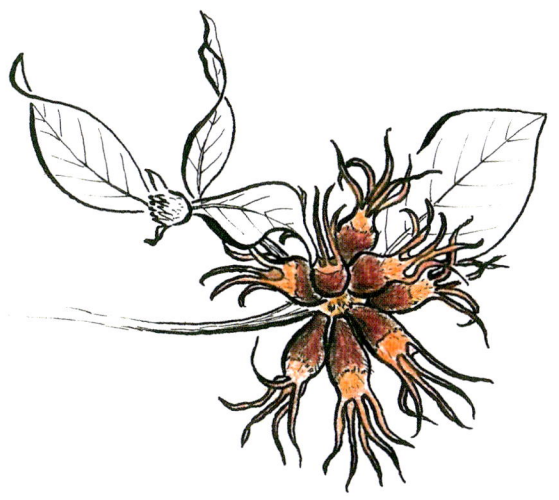

688 | *Burchellia bubalina* | Wild pomegranate | Wildegranaat
Eastern Limpopo, Mpumalanga, KwaZulu-Natal, Eastern and West Cape |
November onwards
Montane forest, forest margins, open moist woodland, grassland, rocky places

A small evergreen tree with bright orange/scarlet flowers in spring

PLATE 363

(75% of actual size)

409 | *Pterocelastrus tricuspidatus* | Candlewood | Kershout (left);
408 | *P. rostratus*| Red candlewood | Rooikershout (right)
Mpumalanga, Eastern Cape, Western Cape; *P. rostratus* absent from East London to George | July to March
P. tricuspidatus: Moist forest margins, dry dune scrub; *P. rostratus*: moist montane and ravine forest

Small- to medium-sized trees that are difficult to identify when not in fruit

PLATE 364

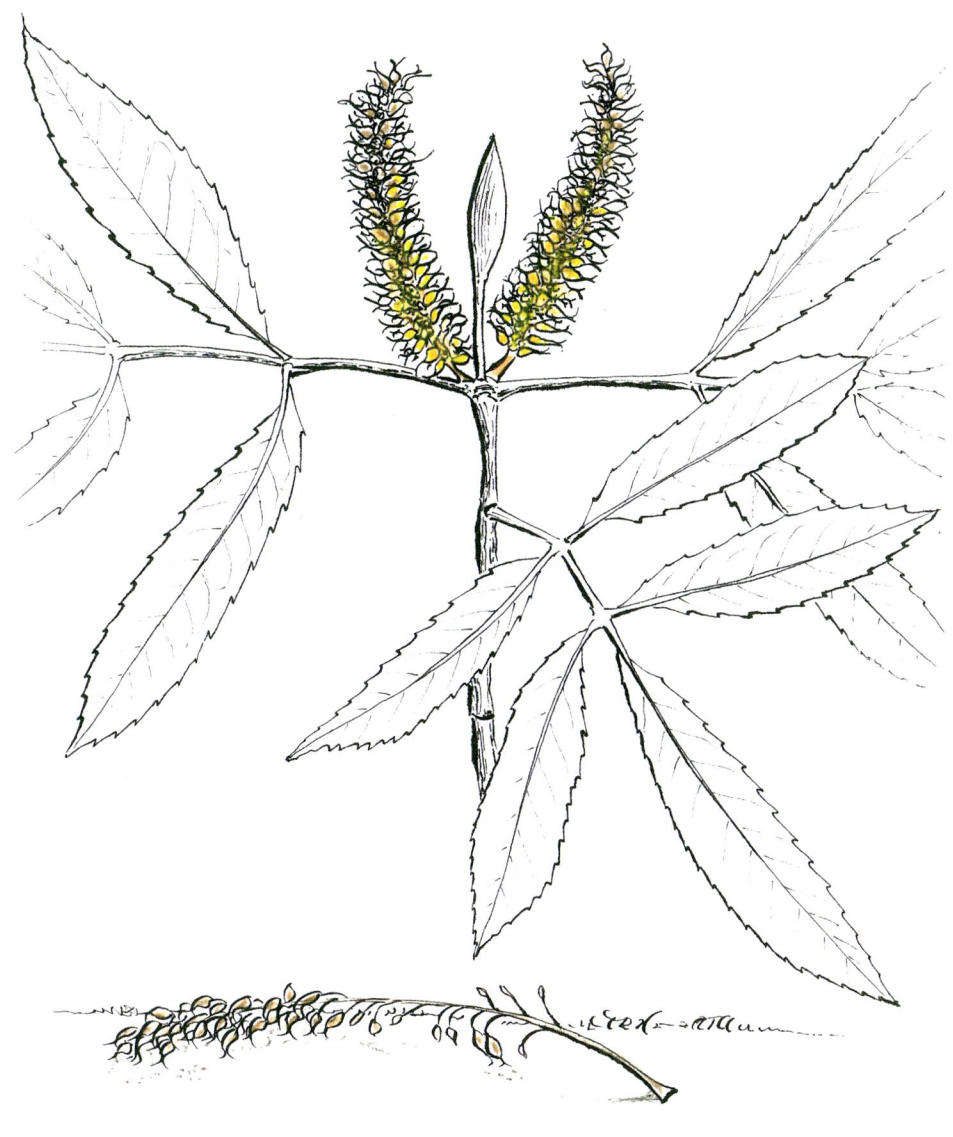

(75% of actual size)

140 | *Cunonia capensis* | Red alder | Rooiels
Eastern KwaZulu-Natal, eastern Eastern Cape, Garden Route to Cape Town | April to July
Moist montane forest, stream banks

A stately evergreen tree; leaf buds covered by two large spoon-shaped stipules

(75% of actual size)

684 | *Breonadia salicina* [=*Adina microcephala*] | Matumi | Mingerhout
Eastern Limpopo, Mpumalanga | June to July
Perennial river banks, stream banks; riverine forest fringes

Tall stately trees, often in small groves with a narrow crown and shiny leaves

PLATE 366

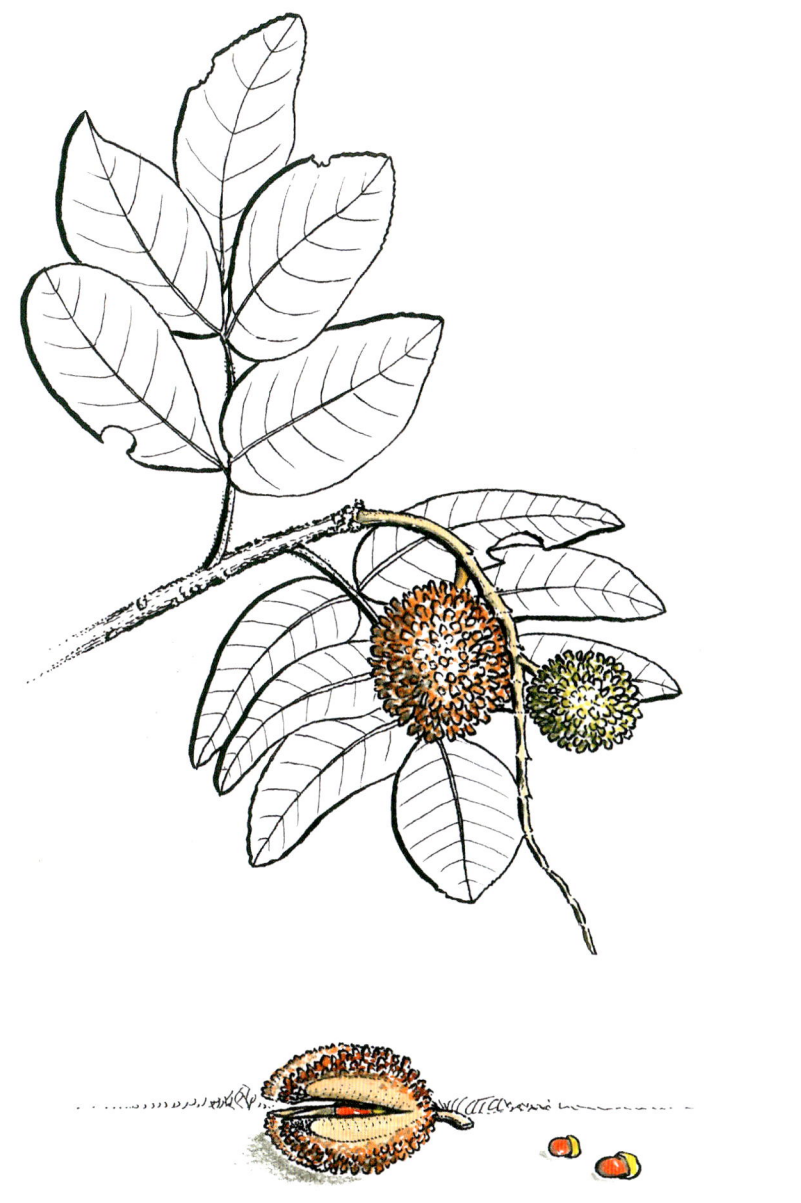

85% of actual size

443 | *Bersama tysoniana* | Common white ash | Gewone witessenhout
Eastern Limpopo, Mpumalanga, coastal KwaZulu-Natal, north-eastern
Eastern Cape | February to August
Margins of coastal and montane forests

Small to large tree; leaves terminally bunched and purplish when young

PLATE 367

(75% of actual size)

441 | *Bersama swinnyi* | Coastal white ash | Kuswitessenhout
Coastal KwaZulu-Natal and north-eastern Eastern Cape | May to July
Coastal evergreen forest and forest margins

A small/medium tree similar to B. tysoniana *(page 383) but with a restricted distribution*

PLATE 368

(40% of actual size)

475 | *Sterculia murex* | Lowveld chestnut | Laeveldkastaiing
Mpumalanga | October to January and later
Bushveld and dry rocky ridges

A medium-sized, well-foliaged tree with attractive bronze spring growth

383

PLATE 369

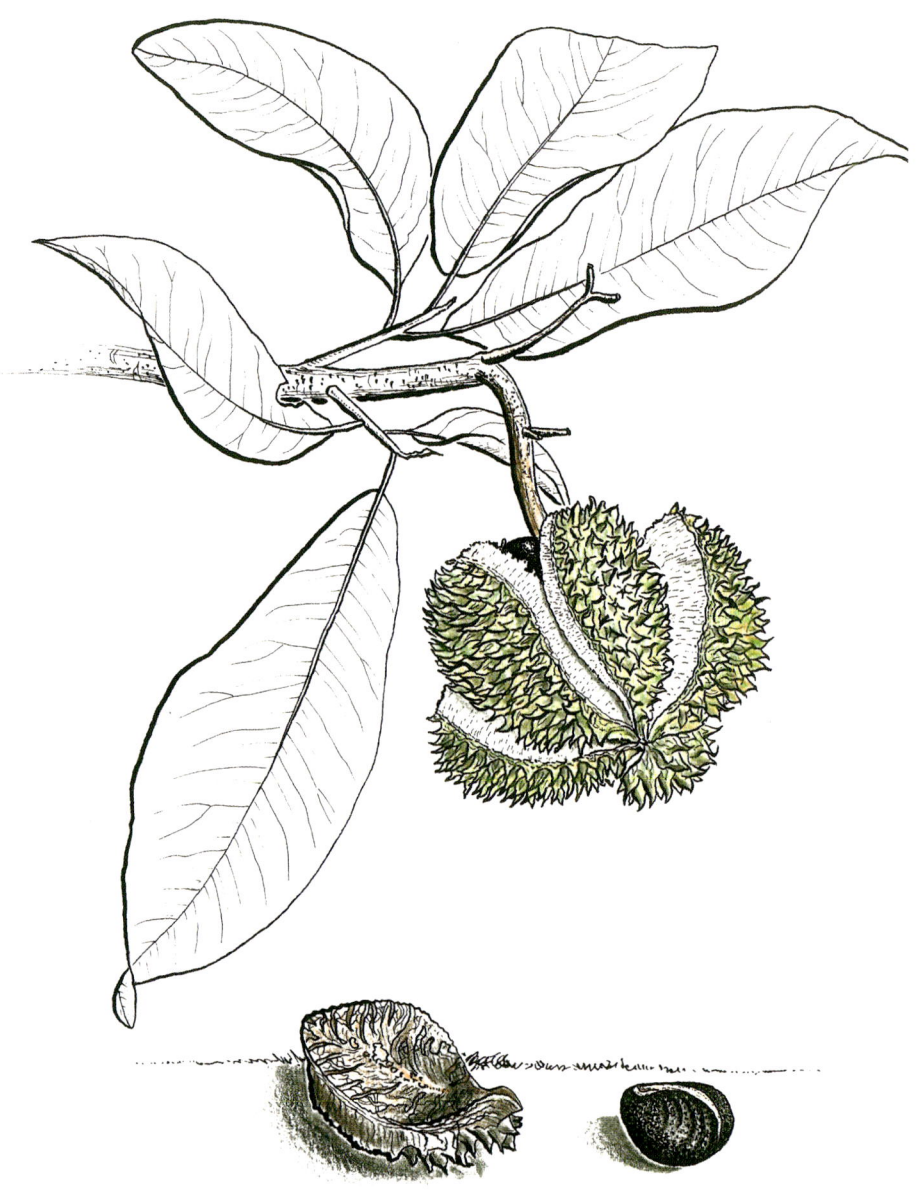

(75% of actual size)

256 | *Calodendrum capense* | Cape chestnut | Wildekastaiing
Eastern Limpopo, Mpumalanga, KwaZulu-Natal, Eastern Cape coast,
Garden Route | January to May
Evergreen Afro-montane forest, wooded ravines, riverine thicket

A forest canopy tree with spectacular pinkish–mauve flowers in spring

PLATE 370

(90% of actual size)

75 | *Faurea saligna* | Willow boekenhout | Treurboekenhout
North West, Limpopo, Gauteng, Mpumalanga, eastern KwaZulu-Natal |
October to April
Open woodland, stony hillsides

A graceful tree with drooping foliage that resembles Eucalyptus *spp. from afar*

PLATE 371

74 | *Faurea macnaughtonii* | Terblans | Terblans
Knysna forest at Gouna; restricted localities in Mpumalanga and KwaZulu-Natal | February to June
Evergreen forests, forest margins

A rare tall forest tree; stem longitudinally striated; lower trunk has flaky bark.

PLATE 372

36.1 | *Salix mucronata* | Cape willow | Kaapse wilger
Widespread except in northern and eastern South Africa and up the Atlantic coast | January to April
Along streams and river banks

A small tree with drooping branches and twigs reddish-purple

PLATE 373

727 | *Brachylaena huillensis* | Lowveld silver oak | Laeveldvaalbos
Northern Limpopo, eastern Mpumalanga, north-eastern KwaZulu-Natal |
October onwards
Dry bushveld, dense scrub, rocky ridges

A small untidy tree, leaves crowded at twig apices, white and woolly below

PLATE 374

20 | *Widdringtonia nodiflora* | Mountain cypress | Bergsipres
Northern Limpopo, western KwaZulu-Natal, Eastern and West Cape |
March onwards
High altitude mountain sides, in rocky gullies

A small upright tree; other **Widdringtonia** *spp. have similar cones.*

PLATE 375

(90% of actual size)

81 | *Leucadendron eucalyptifolium* | Tall conebush | Grootgeelbos
Eastern and South Cape coastal regions | Cones all year round
Coastal mountain forest and fynbos

Shrub or small tree with linear stalkless leaves and leaf base often twisted

(90% of actual size)

77 | *Leucadendron argenteum* | Silver tree | Silwerboom
Table Mountain and Lion's Head | Cones prevalent all year round
Heavy gravel soils derived from granite, Malmesbury shales on mountain slopes

A Cape Peninsula endemic; foliage an unmistakable glistening silvery sheen

PLATE 377

(75% of actual size)

26 | *Raphia australis* | Kosi palm | Kosipalm
Far north-eastern KwaZulu-Natal around Kosi Bay and at Mtunzini |
Once in 25 to 30 years
Swamp forest

A huge palm; leaves 6 to 10 metres long; fruits once after 25 to 30 years, then dies

PLATE 378

607 | *Diospyros natalensis* | Small-leaved jackal-berry | Fynblaarjakkalsbessie
Coastal KwaZulu-Natal and north-eastern Eastern Cape | November to May
Dune forest and scrub, riverine fringes

A small forest tree with small leaves and branches tending to be horizontal

PLATE 379

(75% of actual size)

118 | *Ocotea bullata* | Stinkwood | Stinkhout
Mpumalanga, eastern KwaZulu-Natal, coastal Eastern and West Cape |
March to June
Afro-montane forest

Forest canopy tree; leaves have diagnostic 'bulla' at the base; they look like blisters.

PLATE 380

(75% of actual size)

653 | *Cordia grandicalyx* | Large-fruit saucer-berry | Grootblaarpieringbessie
North-eastern Limpopo | November to December
Hot low altitude bushveld, rocky places

A small tree; loosely spreading crown; biggish circular sandpapery leaves

PLATE 381

(75% of actual size)

25 | *Borassus aethiopum* | Africa fan-palm | Afrikawaaierpalm
Extreme north-eastern Kruger National Park | February to April
Riverine bush and woodland

Large palm; fan-shaped leaves; a prominent swelling halfway up the trunk

About the Author

Trevor Ankiewicz grew up on a Highveld farm in the Bronkhorstspruit District. Having completed secondary school at Belfast High School he furthered his studies at The Saasveld School for Foresters on the outskirts of George where upon graduating in 1966 he was transferred to the South African Forestry Research Institute in Pretoria. There, as a research forester, he was tasked to collect the fruit of ornamental trees in the streets and cemeteries of the capital for the central seed store. He also then worked for several years in the Department of Agricultural Technical Services as a horticulturalist at the Horticultural and Botanical Research Institute in Pretoria. He returned to Saasveld in 1983 and continued working as an extension forester until completing his career in the civil service as a public relations officer for the Department of Nature and Environmental Conservation in the Southern Cape Regional Office in George. Thereafter he became involved as an environmental educator for South African Forestry Company Limited (SAFCOL) with regards to the accreditation by the Forestry Stewardship Council (FSC) of native forests and commercial plantations.

During collecting trips as a pupil forester, he realised the fruit and pods of trees were often an easy way of identification. Here the idea was conceived of producing a field guide concentrating on these diagnostic features to supplement the many excellent publications already available on indigenous South African trees.

Acknowledgements

To say thank you to all the many people and institutions for their support, knowledge and inspiration over a lifetime, which has culminated in this publication, is no easy task and I need to go back a long time in order to include everyone.

- To my parents for making it possible for me to grow up on a delightful Highveld farm and for teaching me the value of trees in general.
- To the patience of my teachers at the Belfast primary and high schools for a firm foundation in my basic education.
- To the Department of Forestry for the training I received at the Saasveld School for Foresters and afterwards.
- To the foresters and scientists of the South African Forestry Research Institute (SAFRI) in Pretoria who acquainted me with sylvicultural practices.
- To the Department of Agricultural Technical Services and the highly professional staff of the Horticultural and Botanical Research Institutes for teaching me so much and for allowing me to work under idyllic conditions.
- To the staff of SABC radio service, as well as Suid Kaap Stereo, for endless hours of delightful entertainment while I toiled at my desk.
- To the scientists, lecturers and technical staff of the Saasveld Research Centre and College for Foresters for always being willing to assist me in my insatiable demands on their time and knowledge and with special thanks to Mr Hora Wilhelmij for his guidance at this time.
- To the staff of the Department of Nature and Environmental Conservation of the Cape Province Administration for a number of years of service and training also under idyllic conditions.

Acknowledgements

- To the directors and staff of Tours for South Africa for giving me the opportunity to collect fruit in so many places in our beautiful country during my tours.
- To my longstanding friend and colleague Wikus van der Walt for his patience and assistance in scanning the illustrations and for providing me with the concept manuscript both electronically and physically.
- To Professor Eugene Moll for his guidance and editing of the manuscript, and for writing the foreword.
- To Dr Ernst van Jaarsveld for his continuous encouragement.
- To Wendy Tate for the concept cover design and introductory layout.
- To my wife Jean (in absentia) for her endless support and encouragement, for being homebuilder and for rearing our son while I pursued my career and worked on this publication.
- To Carol Broomhall of Jacana for supporting my work from the very beginning of my search for a publisher, and to Lara Jacob, Megan Mance and Aimèe Armstrong for taking on the challenge of botanical nomenclature.

Botanical Names Index

Acacia albida – Plate 282	294	*Allophylus decipiens* – Plate 31	34
Acacia ataxacantha – Plate 240	250	*Aloidendron barberae* – Plate 351	366
Acacia burkei – Plate 262	272	*Aloidendron dichotomum* – Plate 351	366
Acacia caffra – Plate 258	268	*Anastrabe integerrima* – Plate 56	59
Acacia erioloba – Plate 278	290	*Androstachys johnsonii* – Plate 312	326
Acacia galpinii – Plate 265	275	*Annona senegalensis* – Plate 219	226
Acacia gerrardii – Plate 275	287	*Anthocleista grandiflora* – Plate 200	205
Acacia grandicornuta – Plate 277	289	*Anthocleista zambesiaca* – Plate 200	205
Acacia haematoxylon – Plate 283	295	*Antidesma venosum* – Plate 22	23
Acacia hebeclada subsp. *hebeclada* – Plate 250	260	*Aphloia theiformis* – Plate 49	52
Acacia hebeclada subsp. *tristis* – Plate 260	270	*Apodytes dimidiata* – Plate 358	373
Acacia karroo – Plate 276	288	*Atalaya alata* – Plate 289	303
Acacia luederitzii var. *retinens* – Plate 249	259	*Avicennia marina* – Plate 150	153
Acacia mellifera – Plate 245	255	*Azanza garckeana* – Plate 193	198
Acacia nigrescens – Plate 259	269	*Balanites maughamii* – Plate 198	203
Acacia nilotica subsp. *kraussiana* – Plate 271	283	*Balanites pedicellaris* – Plate 194	199
Acacia robusta subsp. *robusta* – Plate 255	265	*Barringtonia racemosa* – Plate 221	228
Acacia sieberiana var. *woodii* – Plate 266	276	*Bequaertiodendron magaliesmontanum* – Plate 164	167
Acacia tortilis – Plate 280	292	*Bequaertiodendron natalense* – Plate 170	173
Acacia xanthophloea – Plate 268	280	*Berchemia discolor* – Plate 107	109
Acaica luederitzii var. *luederitzii* – Plate 246	256	*Berchemia zeyheri* – Plate 59	62
Acokanthera oblongifolia –156	159	*Bersama lucens* – Plate 153	156
Acokanthera oppositifolia – Plate 156	159	*Bersama swinnyi* – Plate 367	382
Acokanthera rotundata – Plate 139	142	*Bersama tysoniana* – Plate 366	381
Acokanthera schimperi – Plate 139	142	*Bolusanthus speciosus* – Plate 237	247
Adansonia digitata – Plate 235	244	*Borassus aethiopum* – Plate 381	396
Adina microphala – Plate 365	380	*Boscia albitrunca* – Plate 82	85
Afrocanthium gilfillanii – Plate 318	332	*Boscia mossambicensis* – Plate 127	130
Afrocarpus falcatus – Plate 177	182	*Brabejum stellatifolium* – Plate 206	211
Afzelia quanzensis – Plate 267	277	*Brachylaena huillensis* – Plate 373	388
Alberta magna – Plate 287	301	*Breonadia salicina* – Plate 365	380
Albizia adianthifolia – Plate 256	266	*Bridelia micrantha* – Plate 111	113
Albizia brevifolia – Plate 261	271	*Bridelia mollis* – Plate 47	50
Albizia forbesii – Plate 244	254	*Bruguiera gymnorrhiza* – Plate 329	343
Albizia gummifera – Plate 264	274	*Burchellia bubalina* – Plate 362	377
Albizzia petersiana subsp. *evansii* – Plate 257	267	*Burkea africana* – Plate 247	257
Allophylus africanus – Plate 37	40	*Buxus natalensis* – Plate 171	174

Botanical Names Index

Calodendrum capense – Plate 369	384
Calpurnia aurea – Plate 239	249
Canthium armatum – Plate 112	114
Canthium gilfillanii – Plate 318	332
Canthium inerme – Plate 318	332
Canthium obovatum – Plate 309	323
Capparis tomentosa – Plate 208	215
Carissa macrocarpa – Plate 199	204
Cassia abbreviata – Plate 330	344
Cassine aethiopica – Plate 109	111
Cassine crocea – Plate 143	146
Cassine papillosa – Plate 158	161
Cassine peragua – Plate 35/91	38, 94
Cassine schinoides – Plate 81	84
Cassine transvaalensis – Plate 141	144
Cassinopsis ilicifolia – Plate 106	108
Cassipourea congoensis – Plate 126	128
Cassipourea gerrardii – Plate 126	128
Cassipourea malosana – Plate 126	128
Catha edulis – Plate 341	356
Catunaregam spinosa – Plate 203	208
Catunaregam taylorii – Plate 203	208
Cavacoa aurea – Plate 189	194
Celtis africana – Plate 26	29
Celtis mildbraedii – Plate 80	83
Chaetacme aristata – Plate 104	106
Chionanthus foveolatus – Plate 154	157
Chrysophyllum viridifolium – Plate 209	216
Cleistanthus schlechteri – Plate 313	327
Clerodendrum glabrum – Plate 44	47
Clutia pulchella – Plate 20	21
Cola natalensis – Plate 336	351
Colophospermum mopane – Plate 285	297
Colpoon compressum – Plate 114	116
Combretum celastroides subsp. *orientale* – Plate 298	312
Combretum edwardsii – Plate 300	314
Combretum erythrophyllum – Plate 299	313
Combretum hereroense – Plate 301	315
Combretum imberbe – Plate 302	316
Combretum krausii – Plate 296	310
Combretum molle – Plate 297	311
Combretum zeyheri – Plate 303	317
Commiphora edulis – Plate 162	165
Commiphora glandulosa – Plate 105	107
Commiphora glaucescens – Plate 121	123
Commiphora marlothii – Plate 99/147	101, 150
Commiphora pyracanthoides – Plate 105	107
Cordia caffra – Plate 88	91
Cordia grandicalyx – Plate 380	395
Cordia monoica – Plate 142	145
Cordia ovalis – Plate 142	145
Cordyla africana – Plate 213	220
Crossopteryx febrifuga – Plate 68	71
Croton gratissimus – Plate 151	154
Croton megalobotrys – Plate 324	338
Croton sylvaticus – Plate 123/152	125, 155
Cryptocarya latifolia – Plate 174	179
Cryptocarya myrtifolia – Plate 78	82
Cryptocarya woodii – Plate 133	136
Cryptocarya wyliei – Plate 79	82
Cunonia capensis – Plate 364	379
Curtisia dentata – Plate 86	89
Cussonia spicata – Plate 333	347
Dais cotinifolia – Plate 355	370
Dalbergia melanoxylon – Plate 238	248
Deinbollia oblongifolia – Plate 93	96
Dialium schlechteri – Plate 168	171
Dichrostachys cinerea – Plate 281	293
Diospyros dichrophylla – Plate 183	188
Diospyros glabra – Plate 62	65
Diospyros lycioides – Plate 113/148	115, 151
Diospyros mespiliformis – Plate 134	137
Diospyros natalensis – Plate 378	393
Diospyros scabrida – Plate 357	372
Diospyros whyteana – Plate 346	361
Diplorhynchus condylocarpon – Plate 335	350
Dodenaea viscosa – Plate 295	309
Dodonaea angustifolia – Plate 295	309
Dombeya rotundifolia – Plate 354	369
Dovyalis caffra – Plate 186	191
Dovyalis longispina – Plate 167	170
Dovyalis rhamnoides – Plate 167	170
Dovyalis zeyheri – Plate 169	172
Dracaena aletriformis – Plate 321	335
Dracaena hooeriana – Plate 321	335
Drypetes gerrardii – Plate 317	331
Drypetes natalensis – Plate 185	190
Ekebergia capensis – Plate 145	148
Ekebergia pterophylla – Plate 36	39
Elaeodendron croceum – Plate 158	161
Elaeodendron transvaalensis – Plate 141	144
Elaeodendron zeyheri – Plate 143	146
Empogona lanceolata – Plate 5	6
Englerophytum magaliesmontanum – Plate 164	167
Englerophytum natalense – Plate 170	173
Entandrophragma caudatum – Plate 332	346
Erythrina latissima – Plate 274	286
Erythrina lysistemon – Plate 273	285
Erythrophleum lasianthum – Plate 263	273
Erythrophysa transvaalensis – Plate 347	362
Erythroxylum emarginatum – Plate 110	112
Erythroxylum pictum – Plate 39	42
Euclea crispa – Plate 10	11
Euclea natalensis – Plate 42	45

Euclea pseudebenus – Plate 34	37	*Hippobromus pauciflorus* – Plate 38	41
Euclea racemosa – Plate 46	49	*Homalium dentatum* – Plate 352	367
Euclea undulata – Plate 28/73	31, 78	*Hyaenanche globosa* – Plate 190	195
Eugenia capensis – Plate 144	147	*Hymenodictyon parvifolium* – Plate 205	210
Eugenia erythrophylla – Plate 178	183	*Hyphaene coriacea* – Plate 227	234
Eugenia natalitia – Plate 74	79	*Hyphaene natalensis* – Plate 227	234
Eugenia zeyheri – Plate 131	134	*Ilex mitis* – Plate 4	5
Euphorbia ingens – Plate 100	102	*Indigofera jucunda* – Plate 325	339
Euphorbia tetragona – Plate 311	325	*Inhambanella henriquesii* – Plate 216	223
Euphorbia tirucalli – Plate 124	126	*Kigelia africana* – Plate 334	348
Fagara capensis – Plate 14	15	*Kiggelaria africana* – Plate 192	197
Fagara davyi – Plate 57	60	*Kirkia acuminata* – Plate 343	358
Faidherbia albida – Plate 282	294	*Kirkia wilmsii* – Plate 342	357
Faurea macnaughtonii – Plate 371	386	*Kraussia floribunda* – Plate 45	48
Faurea saligna – Plate 370	385	*Lachnostylis hirta* – Plate 316	330
Ficus abutilifolia – Plate 180	185	*Lannea discolor* – Plate 92	95
Ficus burtt-davyi – Plate 54	57	*Leucadendron argenteum* – Plate 376	391
Ficus capensis – Plate 212	219	*Leucadendron eucalyptifolium* – Plate 375	390
Ficus cordata – Plate 96	98	*Linociera foveolate* – Plate 154	157
Ficus hippopotami – Plate 137	140	*Lonchocarpus capassa* – Plate 248	258
Ficus ilicina – Plate 97	99	*Loxostylis alata* – Plate 356	371
Ficus ingens – Plate 83	86	*Macaranga capensis* – Plate 15	16
Ficus natalensis – Plate 98	100	*Maerua angolensis* – Plate 272	284
Ficus soldanella – Plate 180	185	*Maerua cafra* – Plate 214	221
Ficus sur – Plate 212	219	*Maerua rosmarinoides* – Plate 326	340
Ficus sycomorus – Plate 184	189	*Maesa lanceolata* – Plate 40	43
Ficus trichopoda – Plate 137	140	*Manilkara concolor* – Plate 117	119
Flacourtia indica – Plate 207	214	*Manilkara discolor* – Plate 116	118
Galpinia transvaalica – Plate 16	17	*Margaritaria discoidea* – Plate 310	324
Garcinia gerrardii – Plate 175	180	*Markhamia acuminata* – Plate 284	296
Garcinia livingstonei – Plate 201	206	*Markhamia zanzibarica* – Plate 284	296
Gardenia cornuta – Plate 226	233	*Maurocenia frangularia* – Plate 84	87
Gardenia spatulifolia – Plate 234	243	*Maytenus acuminata* – Plate 69/315	72, 329
Gardenia thunbergia – Plate 230	238	*Maytenus heterophylla* – Plate 18/72	19, 75
Gardenia volkensii – Plate 234	243	*Maytenus oleoides* – Plate 67	70
Gerrardina foliosa – Plate 25	28	*Maytenus peduncularis* – Plate 66	69
Gonioma kamassi – Plate 327	341	*Maytenus procumbens* – Plate 308	322
Grewia flavescens – Plate 108	110	*Maytenus undata* – Plate 101	103
Grewia hexamita – Plate 322	336	*Memecylon bachmannii* – Plate 75	80
Grewia monticola – Plate 305	319	*Memecylon grandiflorum* – Plate 75	80
Greyia sutherlandii – Plate 344	359	*Memecylon natalense* – Plate 76	80
Gymnosporia buxifolia – Plate 18/72	19, 75	*Millettia grandis* – Plate 254	264
Gyrocarpus americanus – Plate 288	302	*Mimusops caffra* – Plate 157	160
Halleria lucida – Plate 85	88	*Mimusops zeyheri* – Plate 182	187
Harpephyllum caffrum – Plate 160	163	*Monodora junodii* – Plate 217	224
Hartogia schinoides – Plate 81	84	*Mundulea sericea* – Plate 241	251
Heeria argentea – Plate 187	192	*Mystroxylon aethiopicum* – Plate 109	111
Heteromorpha arborescens – Plate 70	73	*Notobuxus natalensis* – Plate 171	174
Heteromorpha trifoliata – Plate 70	73	*Nuxia floribunda* – Plate 24	25
Heteropyxis natalensis – Plate 19	20	*Nymania capensis* – Plate 345	360
Hexalobus monopetalus – Plate 328	342	*Ochna arborea* – Plate 359	374
Heywoodia lucens – Plate 320	334	*Ochna pulchra* – Plate 360	375

Botanical Names Index

Ocotea bullata – Plate 379	394	*Rapanea melanophloeos* – Plate 51	54
Olax dissitiflora – Plate 63	66	*Raphia australis* – Plate 377	392
Olea africana – Plate 64	67	*Rauvolfia caffra* – Plate 138	141
Olea capensis subsp. *capensis* – Plate 163	166	*Rawsonia lucida* – Plate 223	230
Olea capensis subsp. *macrocarpa* – Plate 163	166	*Rhamnus prinoides* – Plate 27	30
Olea europaea subsp. *africana* – Plate 64	67	*Rhizophora mucronata* – Plate 329	343
Olea exasperata – Plate 64	67	*Rhus chirindensis* – Plate 3	4
Olinia emarginata – Plate 94	97	*Rhus dentata* – Plate 29	32
Olinia ventosa – Plate 95	97	*Rhus lancea* – Plate 11	12
Oncoba spinosa – Plate 228	236	*Rhus leptodictya* – Plate 12	13
Oricia bachmannii – Plate 165	168	*Rhus penduline* – Plate 6	7
Osyris compressa – Plate 114	116	*Rothmannia capensis* – Plate 231	240
Osyris lanceolata – Plate 103	105	*Rothmannia fischeri* – Plate 220	227
Osyris quadripartita – Plate 103	105	*Rothmannia globosa* – Plate 188	193
Oxyanthus natalensis – Plate 197	202	*Salix mucronata* – Plate 372	387
Oxyanthus pyriformis – Plate 197	202	*Salvadora australis* – Plate 55	58
Ozoroa dispar – Plate 120	122	*Salvadora australis* – Plate 55	58
Ozoroa paniculosa – Plate 65	68	*Sapium ellipticum* – Plate 307	321
Ozoroa sphaerocarpa – Plate 65	68	*Sapium integerrimum* – Plate 323	337
Pancovia golungensis – Plate 202	207	*Sapium integerrimus* – Plate 323	337
Pappea capensis – Plate 172	175	*Schefflera umbellifera* – Plate 9	10
Parinari curatellifolia – Plate 218	225	*Schotia afra* var. *afra* – Plate 252	262
Parkinsonia africana – Plate 269	281	*Schotia brachypetala* – Plate 251	261
Pavetta edentula – Plate 52	55	*Schotia latifolia* – Plate 253	263
Pavetta eylesii – Plate 48	51	*Schrebera alata* – Plate 204	209
Pavetta lanceolata – Plate 32	35	*Sclerocarya birrea* – Plate 211	218
Peddiea africana – Plate 115	117	*Sclerocarya caffra* – Plate 211	218
Peltophorum africanum – Plate 236	246	*Sclerocroton integerrimus* – Plate 323	337
Philenoptera violacea – Plate 248	258	*Scolopia mundii* – Plate 33/60	36, 63
Phoenix reclinata – Plate 118	120	*Scolopia zeyheri* – Plate 21	22
Phylica paniculata – Plate 13	14	*Searsia chirindensis* – Plate 3	4
Piliostigma thonningii – Plate 279	291	*Searsia dentata* – Plate 29	32
Pittosporum viridiflorum – Plate 306	320	*Searsia lancea* – Plate 11	12
Platylophus trifoliatus – Plate 348	363	*Searsia leptodictya* – Plate 12	13
Plectroniella armata – Plate 112	114	*Searsia lucida* – Plate 2	3
Pleurostylia capensis – Plate 155	158	*Searsia pendulina* – Plate 6	7
Podocarpus elongatus – Plate 361	376	*Searsia pyroides* – Plate 1	2
Podocarpus falcatus – Plate 177	182	*Securidaca longipedunculata* – Plate 286	300
Podocarpus henkelii – Plate 135	138	*Senegalia ataxacantha* – Plate 240	250
Podocarpus latifolius – Plate 361	376	*Senegalia burkei* – Plate 262	272
Premna mooiensis – Plate 8	9	*Senegalia caffra* – Plate 258	268
Protorhus longifolia – Plate 119	121	*Senegalia galpinii* – Plate 265	275
Prunus africana – Plate 30	33	*Senegalia mellifera* – Plate 245	255
Pseudolachnostylis maprouneifolia – Plate 173	178	*Senegalia nigrescens* – Plate 259	269
Psychotria capensis – Plate 50	53	*Shirakiopsis elliptica* – Plate 307	321
Psydrax obovata – Plate 309	323	*Sideroxylon inerme* – Plate 87	90
Ptaeroxylon obliquum – Plate 349	364	*Smelophyllum capense* – Plate 319	333
Pterocarpus angolensis – Plate 291	305	*Solanum giganteum* – Plate 53	56
Pterocarpus rotundifolius – Plate 290	304	*Spirostachys africana* – Plate 304	318
Pterocelastrus rostratus – Plate 363	378	*Steganotaenia araliacea* – Plate 350	365
Pterocelastrus tricuspidatus – Plate 363	378	*Sterculia murex* – Plate 368	383
Putterlickia pyracantha – Plate 125	127	*Sterculia rogersii* – Plate 339	354

Strychnos cocculoides – Plate 233	242	*Vachellia hebeclada* subsp. *hebeclada* – Plate 250	260
Strychnos decussata – Plate 146	149	*Vachellia hebeclada* subsp. *tristis* – Plate 260	270
Strychnos henningsii – Plate 140	143	*Vachellia karroo* – Plate 276	288
Strychnos madagascariensis – Plate 229	237	*Vachellia luederitzii* var. *luederitzii* – Plate 246	256
Strychnos mitis – Plate 89	92	*Vachellia luederitzii* var. *retinens* – Plate 249	259
Strychnos pungens – Plate 232	241	*Vachellia nilotica* subsp. *kraussiana* – Plate 271	283
Strychnos usambarensis – Plate 132	135	*Vachellia robusta* subsp. *robusta* – Plate 255	265
Suregada africana – Plate 314	328	*Vachellia sieberiana* var. *woodii* – Plate 266	276
Syzygium cordatum – Plate 102	104	*Vachellia tortilis* – Plate 280	292
Syzygium guineense – Plate 161	164	*Vachellia xanthophloea* – Plate 268	280
Tabernaemontana elegans – Plate 338	353	*Vangueria cyanescens* – Plate 181	186
Tabernaemontana ventricosa – Plate 337	352	*Vangueria infausta* – Plate 179	184
Tapiphyllum parvifolium – Plate 129	132	*Vangueria parvifolia* – Plate 129	132
Tarenna pavettoides – Plate 43	46	*Vepris lanceolata* – Plate 41	44
Teclea gerrardii – Plate 122	124	*Vepris reflexa* – Plate 90	93
Teclea natalensis – Plate 61	64	*Vepris undulate* – Plate 41	44
Terminalia phanerophlebia – Plate 293	307	*Virgilia oroboides* – Plate 242	252
Terminalia prunioides – Plate 292	306	*Vitellariopsis marginata* – Plate 215	222
Terminalia sericea – Plate 294	308	*Vitex harveyana* – Plate 77	81
Thespesia acutiloba – Plate 128	131	*Vitex mombassae* – Plate 176	181
Toddaliopsis bremekampii – Plate 149	152	*Vitex obovate* subsp. *wilmsii* – Plate 353	368
Trema orientalis – Plate 7	8	*Vitex wilmsii* – Plate 353	368
Tricalysia capensis – Plate 58	61	*Voacanga thouarsii* – Plate 340	355
Tricalysia lanceolata – Plate 5	6	*Warburgia salutaris* – Plate 210	217
Trichilia dregeana – Plate 224	231	*Widdringtonia nodiflora* – Plate 374	389
Trichilia emetica – Plate 225	232	*Wrightia natalensis* – Plate 331	345
Trichocladus grandiflorus – Plate 17	18	*Xanthocercis zambesiaca* – Plate 195	200
Trimeria grandifolia – Plate 23	24	*Ximenia americana* – Plate 130	133
Trimeria rotundifolia – Plate 23	24	*Ximenia caffra* – Plate 159/196	162, 201
Turraea floribunda – Plate 191	196	*Xuromphis obovate* – Plate 203	208
Turraea nilotica – Plate 71	74	*Xylopia parviflora* – Plate 270	282
Umtiza listeriana – Plate 243	253	*Xylotheca kraussiana* – Plate 222	229
Vachellia erioloba – Plate 278	290	*Xymalos monospora* – Plate 166	169
Vachellia gerrardii – Plate 275	287	*Zanthoxylum capense* – Plate 14	15
Vachellia grandicornuta – Plate 277	289	*Zanthoxylum davyi* – Plate 57	60
Vachellia haematoxylon – Plate 283	295	*Ziziphus mucronata* – Plate 136	139

Most Common English Names

African dog rose – Plate 222	229	Brown ivory – Plate 107	109
African fan palm – Plate 381	396	Buffalo thorn – Plate 136	139
African mangosteen – Plate 201	206	Bush boer-bean – Plate 253	263
African wattle – Plate 236	246	Bush medlar – Plate 181	186
Albino berry – Plate 49	52	Bushman's tea – Plate 341	356
Ana tree – Plate 282	294	Bushveld honeysuckle tree – Plate 71	74
Apple leaf – Plate 248	258	Bushveld red balloon – Plate 347	362
Assegai – Plate 86	89	Bushveld saffron – Plate 141	144
Balloon thorn – Plate 249	259	Bushveld white ironwood – Plate 90	93
Baobab – Plate 235	244	Cabbage tree – Plate 333	347
Bead-bean tree – Plate 272	284	Camel thorn – Plate 278	290
Bean tree – Plate 284	296	Camel's foot – Plate 279	291
Bell gardenia – Plate 188	193	Candelabra tree – Plate 100	102
Black bird-berry – Plate 50	53	Candle acacia – Plate 250	260
Black false currant – Plate 37	40	Candle thorn – Plate 260	270
Black mangrove – Plate 329	343	Candlewood – Plate 363	378
Black monkey orange – Plate 229	237	Cape ash – Plate 145	148
Black monkey thorn – Plate 262	272	Cape beech – Plate 51	54
Black thorn – Plate 245	255	Cape blackwood – Plate 66	69
Bladder-nut – Plate 346	361	Cape chestnut – Plate 369	384
Blossom tree – Plate 242	252	Cape coffee – Plate 58	61
Blue bitter berry – Plate 132	135	Cape gardenia – Plate 231	240
Blue bush – Plate 113/148	115, 151	Cape holly – Plate 4	5
Blue guarri – Plate 10	11	Cape plane – Plate 359	374
Blue-leaved corkwood – Plate 121	123	Cape quince – Plate 133	136
Blueberry bush – Plate 62	65	Cape saffron – Plate 35/91	38, 94
Bluebush (Spioenkop) – Plate 113/148	115, 151	Cape teak – Plate 146	149
Blue sourplum – Plate 130	133	Cape willow – Plate 372	387
Breede river yellowwood – Plate 361	376	Carrot tree – Plate 350	365
Brittle bush – Plate 319	333	Cheesewood – Plate 306	320
Broad-pod albizzia – Plate 244	254	Chinese lantern – Plate 345	360
Broad-leaved bride's bush – Plate 48	51	Coalwood – Plate 316	330
Broad-leaved coral tree – Plate 274	286	Coastal bladder-nut – Plate 357	372
Broad-leaved quince – Plate 174	179	Coastal red milkwood – Plate 157	160
Broad-leaved shepherd's tree – Plate 127	130	Coastal tannin-bush – Plate 103	105
Broom cluster fig – Plate 212	219	Coastal white ash – Plate 367	382
Brown ironwood – Plate 352	367	Coffee pear – Plate 155	158

Know them by their Fruit

Entry	Page
Coffee-bean strychnos – Plate 140	143
Common bush-cherry – Plate 214	221
Common canary-berry – Plate 314	328
Common coca-tree – Plate 110	112
Common cola – Plate 336	351
Common coral tree – Plate 273	285
Common corkwood – Plate 105	107
Common forest myrtle – Plate 74	79
Common guarri – Plate 28/73	31, 78
Common hard-leaf – Plate 13	14
Common hook thorn – Plate 258	268
Common onionwood – Plate 126	128
Common poison bush – Plate 156	159
Common resin tree – Plate 65	68
Common saffron – Plate 158	161
Common sourberry – Plate 167	170
Common spike-thorn – Plate 18/72	19, 75
Common star-apple – Plate 183	188
Common star-chestnut – Plate 339	354
Common tree euphorbia – Plate 100	102
Common wild currant – Plate 1	2
Common white ash – Plate 366	381
Cork bush – Plate 241	251
Corky-bark monkey orange – Plate 233	242
Crystal bark – Plate 68	71
Currant resin tree – Plate 65	68
Dogwood – Plate 27	30
Duiker berry – Plate 323	337
Dune coco tree – Plate 308	322
Dune myrtle – Plate 144	147
Dune olive – Plate 64	67
Dune poison bush – Plate 64	67
Dune soap-berry – Plate 93	96
Ebony tree – Plate 34	37
Eastern aloe-tree – Plate 351	366
False assegai – Plate 40	43
False bride's bush – Plate 43	46
False cabbage tree – Plate 9	10
False currant – Plate 31	34
False horsewood – Plate 38	41
False soap berry – Plate 202	207
False spike thorn – Plate 125	127
False tamboti – Plate 313	327
False turkey berry – Plate 112	114
False umbrella thorn – Plate 246	256
Fever tree – Plate 268	280
Flame thorn – Plate 240	250
Flat crown – Plate 256	266
Fluted milkwood – Plate 209	216
Forest bride's bush – Plate 32	35
Forest bushwillow – Plate 296	310
Forest coca tree – Plate 39	42
Forest croton – Plate 123/152	125, 155
Forest elder – Plate 24	25
Forest fever tree – Plate 200	205
Forest iron plum – Plate 317	331
Forest mahogany – Plate 224	231
Forest mangosteen – Plate 175	180
Forest milkberry – Plate 116	118
Forest peach – Plate 223	230
Forest toad tree – Plate 337	352
Giant raisin – Plate 322	336
Glossy currant – Plate 2	3
Glossy bersama – Plate 153	156
Governor's plum – Plate 207	214
Green apple – Plate 217	224
Green witch hazel – Plate 17	18
Grey camel thorn – Plate 283	295
Hairy fingerleaf – Plate 353	368
Hard pear – Plate 95	97
Henkel's yellowwood – Plate 135	138
Honey euphorbia – Plate 311	325
Horn-pod tree – Plate 335	350
Horned thorn – Plate 277	289
Hottentot's cherry – Plate 84	87
Hyaena-poison – Plate 190	195
Ilala palm – Plate 227	234
Ironwood – Plate 163	166
Jackal berry – Plate 134	137
Jackal coffee – Plate 5	6
Jacket plum – Plate 172	175
Jumping-seed tree – Plate 307	321
Kamassi – Plate 327	341
Karee – Plate 11	12
Karoo boer-bean – Plate 252	262
Kei-apple – Plate 186	191
Knob thorn – Plate 259	269
Knobwood – Plate 57	60
Koko tree – Plate 101	103
Kooboo berry – Plate 109	111
Kosi palm – Plate 377	392
Krantz berry – Plate 25	28
Kudu berry – Plate 173	178
Large fever-berry – Plate 324	338
Large num-num – Plate 199	204
Large-fruit saucer-berry – Plate 380	395
Large-fruited bushwillow – Plate 303	317
Large-leaved bride's bush – Plate 52	55
Large-leaved dragon tree – Plate 321	335
Large-leaved myrtle – Plate 178	183
Large-leaved rock fig – Plate 180	185
Large sour plum – Plate 159/196	162, 201
Laurel fig – Plate 97	99
Lavender croton – Plate 151	154

Most Common English Names

Name	Page
Lavender tree – Plate 19	20
Leadwood – Plate 302	316
Lebombo cluster-leaf – Plate 293	307
Lebombo ironwood – Plate 312	326
Lebombo wing-nut – Plate 289	303
Lemon thorn – Plate 106	108
Lemon wood – Plate 166	169
Live-long – Plate 92	95
Lowveld chestnut – Plate 368	383
Lowveld cluster-leaf – Plate 292	306
Lowveld silver oak – Plate 373	388
Many-stemmed albizia – Plate 257	267
Marula – Plate 211	218
Matumi – Plate 365	380
Milk pear – Plate 216	223
Mitzeeri – Plate 111	113
Mobola plum – Plate 218	225
Monkey thorn – Plate 265	275
Mopane – Plate 285	297
Mountain albizia – Plate 261	271
Mountain cypress – Plate 374	389
Mountain hard-pear – Plate 94	97
Mountain karee – Plate 12	13
Mountain medlar – Plate 129	132
Mountain seringa – Plate 342	357
Myrtle quince – Plate 78	82
Namaqua fig – Plate 96	98
Namaqua resin tree – Plate 120	122
Nana berry – Plate 29	32
Narrow-leaved mustard tree – Plate 55	58
Natal apricot – Plate 167	170
Natal bottlebrush – Plate 344	359
Natal box – Plate 171	174
Natal bush milkwood – Plate 215	222
Natal cherry-orange – Plate 61	64
Natal combretum – Plate 300	314
Natal fig – Plate 98	100
Natal flame bush – Plate 287	301
Natal gardenia – Plate 226	233
Natal guarri – Plate 42	45
Natal hickory – Plate 189	194
Natal ironplum – Plate 185	190
Natal loquat – Plate 197	202
Natal mahogany – Plate 225	232
Natal rose apple – Plate 76	80
Natal white stinkwood – Plate 80	83
Needle-leaved bush cherry – Plate 326	340
Nyala tree – Plate 195	200
Outeniqua yellowwood – Plate 177	182
Pambati tree – Plate 56	59
Paperbark corkwood – Plate 99/147	101, 150
Paperbark thorn – Plate 266	276
Parsley tree – Plate 70	73
Peeling plane – Plate 360	375
Pepper-bark tree – Plate 210	217
Pheasant-berry – Plate 310	324
Pigeonwood – Plate 7	8
Pock ironwood – Plate 154	157
Pod mahogany – Plate 267	277
Poison olive – Plate 115	117
Pompom tree – Plate 355	370
Pondo rose apple – Plate 75	80
Poora berry – Plate 176	181
Powder puff tree – Plate 221	228
Propeller tree – Plate 288	302
Quar – Plate 309	323
Quinine tree – Plate 138	141
Quiver tree – Plate 351	366
Real yellowwood – Plate 361	376
Red alder – Plate 364	379
Red beech – Plate 119	121
Red bitter apple – Plate 53	56
Red candlewood – Plate 363	378
Red currant – Plate 3	4
Red ivory – Plate 59	62
Red mangrove – Plate 329	343
Red milkwood – Plate 182	187
Red pear – Plate 33/60	36, 63
Red quince – Plate 79	82
Red stinkwood – Plate 30	33
Red thorn – Plate 275	287
Red-leaved rock fig – Plate 83	86
Rhino coffee – Plate 45	48
River bushwillow – Plate 299	313
River indigo – Plate 325	339
River thorn – Plate 255	265
Rock ash – Plate 36	39
Rock candlewood – Plate 67	70
Rockwood – Plate 187	192
Rough-leaved corkwood – Plate 162	165
Rough-leaved raisin – Plate 108	110
Round-leaved kiaat – Plate 290	304
Round-leaved poison bush – Plate 139	142
Rubber hedge euphorbia – Plate 124	126
Russet bushwillow – Plate 301	315
Saddle pod – Plate 331	345
Sand olive – Plate 295	309
Satin-bark saucer-bush – Plate 142	145
Sausage tree – Plate 334	348
Savanna bushwillow – Plate 298	312
Savanna gardenia – Plate 234	243
Scented thorn – Plate 271	283
Sea guarri – Plate 46	49
Septee saucer-berry – Plate 88	91

Name	Page
Shakama plum – Plate 328	342
Shepherd's tree – Plate 82	85
Sickle bush – Plate 281	293
Silky bark – Plate 69/315	72, 329
Silver cluster-leaf – Plate 294	308
Silver-leaved milkplum – Plate 170	173
Silver raisin – Plate 305	319
Silver tree – Plate 376	391
Sjambok pod – Plate 330	344
Skunk bush – Plate 8	9
Small ironwood – Plate 163	166
Small knobwood – Plate 14	15
Small sourplum – Plate 63	66
Small torchwood – Plate 194	199
Small-leaved jackal-berry – Plate 378	393
Small-leaved saffron – Plate 143	146
Smooth-bark albizia – Plate 264	274
Sneezewood – Plate 349	364
Snot apple – Plate 193	198
Snuff-box tree – Plate 228	236
Spine-leaved monkey orange – Plate 232	241
Spoonwood – Plate 81	84
Stamvrug milkplum – Plate 164	167
Stink ebony – Plate 320	334
Stinkwood – Plate 379	394
Swamp fig – Plate 137	140
Swazi ordeal tree – Plate 263	273
Sweet thorn – Plate 276	288
Sycamore fig – Plate 184	189
Tall bitterwood – Plate 270	282
Tall conebush – Plate 375	390
Tannin bush – Plate 103	105
Tamboti – Plate 304	318
Tarwood – Plate 356	371
Tassel berry – Plate 22	23
Terblans – Plate 371	386
Thorn pear – Plate 21	22
Thorny bone-apple – Plate 203	208
Thorny elm – Plate 104	106
Three-finger vitex – Plate 77	81
Tinderwood – Plate 44	47
Toad tree – Plate 338	353
Torchwood – Plate 198	203
Tree fuchsia – Plate 85	88
Tree wisteria – Plate 237	247
Turkey berry – Plate 318	332
Twin-berry tree – Plate 165	168
Umbrella thorn – Plate 280	292
Umtiza – Plate 243	253
Umzimbeet – Plate 254	264
Veld fig – Plate 54	57
Velvet bushwillow – Plate 297	311
Velvet rock-alder – Plate 318	332
Velvet sweetberry – Plate 47	50
Violet tree – Plate 286	300
Warty-fruited lightning bush – Plate 20	21
Water berry – Plate 102	104
Weeping boer-bean – Plate 251	261
White alder – Plate 348	363
White gardenia – Plate 230	238
White ironwood – Plate 41	44
White karee – Plate 6	7
White mangrove – Plate 150	153
White milkwood – Plate 87	90
White pear – Plate 358	373
White seringa – Plate 343	358
White stinkwood – Plate 26	29
Wild almond – Plate 206	211
Wild apricot – Plate 169	172
Wild custard apple – Plate 219	226
Wild date palm – Plate 118	120
Wild frangipani – Plate 340	355
Wild green-hair tree – Plate 269	281
Wild honeysuckle tree – Plate 191	196
Wild laburnum – Plate 239	249
Wild mandarin – Plate 149	152
Wild mango – Plate 213	220
Wild medlar – Plate 179	184
Wild mulberry – Plate 23	24
Wild myrtle – Plate 131	134
Wild olive – Plate 64	67
Wild peach – Plate 192	197
Wild pear – Plate 354	369
Wild plum – Plate 160	163
Wild pomegranate – Plate 362	377
Wild poplar – Plate 15	16
Wild pride of India – Plate 16	17
Wild seringa – Plate 247	257
Wild teak – Plate 291	305
Wild tulip tree – Plate 128	131
Willow boekenhout – Plate 370	385
Wing-leaved wooden pear – Plate 204	209
Wooden banana – Plate 332	346
Woodland gardenia – Plate 220	227
Woodland waterberry – Plate 161	164
Woolly caper-bush – Plate 208	215
Yellow bitterberry – Plate 89	92
Yellow firebush – Plate 205	210
Zebra wood – Plate 238	248
Zulu cherry-orange – Plate 122	124
Zulu milkberry – Plate 117	119
Zulu podberry – Plate 168	171

Most Common Afrikaans Names

Name	Page
Afrika hondsroos – Plate 222	229
Afrikawaaierpalm – Plate 381	396
Anaboom – Plate 282	294
Apiesdoring – Plate 265	275
Appelblaar – Plate 248	258
Assegaai – Plate 86	89
Basboom – Plate 355	370
Basterassegai – Plate 40	43
Baster-haak-en-steek – Plate 246	256
Basterbokdrol – Plate 112	114
Basterbruidsbos – Plate 43	46
Basterkiepersol – Plate 9	10
Basterpendoring – Plate 125	127
Basterperdepis – Plate 38	41
Bastersaffraan – Plate 35/91	38, 94
Basterseepbessie – Plate 202	207
Bastersuurpruim – Plate 63	66
Bastertaaibos – Plate 31	34
Bastertambotie – Plate 313	327
Basterwitysterhout – Plate 90	93
Bergbas – Plate 103	105
Berghardepeer – Plate 94	97
Bergkaree – Plate 12	13
Bergmahonie – Plate 332	346
Bergmispel – Plate 129	132
Bergperske – Plate 49	52
Bergsering – Plate 342	357
Bergsipres – Plate 374	389
Bergvalsdoring – Plate 261	271
Besemtrosvy – Plate 212	219
Blaasdoring – Plate 249	259
Blinktaaibos – Plate 2	3
Blinkblaar – Plate 27	30
Blinkblaar-wag-'n-bietjie – Plate 136	139
Blinkblaarmispel – Plate 181	186
Blinkblaarwitessenhout – Plate 153	156
Bloubessiebos – Plate 62	65
Bloubitterbessie – Plate 132	135
Bloublaarkanniedood – Plate 121	123
Bloubos – Plate 113/148	115, 151
Bloughwarrie – Plate 10	11
Blousuurpruim – Plate 130	133
Boesmanstee – Plate 341	356
Boomaalwyn – Plate 351	366
Bosboerboon – Plate 253	263
Bosbruidsbos – Plate 32	35
Bosgeelmelkhout – Plate 175	180
Boskokaboom – Plate 39	42
Boskoorsbessie – Plate 123/152	125, 155
Boskoorsboom – Plate 200	205
Bosmelkbessie – Plate 116	118
Bospaddaboom – Plate 337	352
Bosperske – Plate 223	230
Bosrooiessenhout – Plate 224	231
Bosstamvrug – Plate 209	216
Bostaaibos – Plate 3	4
Bostolbos – Plate 346	361
Bosvaderlandswilg – Plate 296	310
Bosveldkamperfoelieboom – Plate 71	74
Bosveldkatjiepiering – Plate 234	243
Bosveldsaffraan – Plate 141	144
Bosveld liguster – Plate 16	17
Bosveldrooiklapperbos – Plate 347	362
Bosveldwitklokke – Plate 220	227
Bosvlier – Plate 24	25
Bosysterpruim – Plate 317	331
Breëblaarbruidsbos – Plate 48	51
Breëblaarkoraalboom – Plate 274	286
Breëblaarkweper – Plate 174	179
Breëblaarwitgat – Plate 127	130
Breekhout – Plate 287	301
Breëpeulvalsdoring – Plate 244	254
Breëriviergeelhout – Plate 361	376
Bruinivoor – Plate 107	109
Bruinysterhout – Plate 352	367
Buig-my-nie – Plate 319	333

409

Name	Page
Plate 92	95
˙kiaat – Plate 290	304
ˌpruim – Plate 172	175
ˌringbeenappel – Plate 203	208
Doringolm – Plate 104	106
Doringpeer – Plate 21	22
Drolpeer – Plate 354	369
Duikerbessie – Plate 323	337
Duinegifboom – Plate 156	159
Duinekokoboom – Plate 308	322
Duinemirt – Plate 144	147
Duineolien – Plate 64	67
Duineseepbessie – Plate 93	96
Ebbeboom – Plate 34	37
Enkeldoring – Plate 255	265
Essenhout – Plate 145	148
Fisantebessie – Plate 310	324
Fluweelboswilg – Plate 297	311
Fluweelklipels – Plate 318	332
Fluweelsoetbessie – Plate 47	50
Fynblaarjakkalsbessie – Plate 378	393
Fynblaarsaffraan – Plate 143	146
Geelbitterbessie – Plate 89	92
Geelbrandbos – Plate 205	210
Geelklapper – Plate 233	242
Geelwortelboom – Plate 350	365
Gewone bliksembos – Plate 20	21
Gewone bokdrol – Plate 318	332
Gewone ghwarrie – Plate 28/73	31, 78
Gewone gifboom – Plate 156	159
Gewone haakdoring – Plate 258	268
Gewone hardeblaar – Plate 13	14
Gewone harpuisboom – Plate 65	68
Gewone kanariebessie – Plate 314	328
Gewone kanniedood – Plate 105	107
Gewone kiepersol – Plate 333	347
Gewone kokaboom – Plate 110	112
Gewone koraalboom – Plate 273	285
Gewone naboom – Plate 100	102
Gewone pendoring – Plate 18/72	19, 75
Gewone saffraan – Plate 158	161
Gewone sterappel – Plate 183	188
Gewone sterkastaiing – Plate 339	354
Gewone suurbessie – Plate 167	xx
Gewone taaibos – Plate 1	2
Gewone trosvy – Plate 184	189
Gewone uiehout – Plate 126	128
Gewone witbos – Plate 214	221
Gewone witessenhout – Plate 366	381
Gifolyf – Plate 115	117
Goewerneurspruim – Plate 207	214
Groen towerhaselaar – Plate 17	18
Groenappel – Plate 217	224
Groendoring – Plate 198	203
Grootnoemnoem – Plate 199	204
Grootbitterappel – Plate 53	56
Grootbitterhout – Plate 270	282
Grootblaarbruidsbos – Plate 52	55
Grootblaardrakeboom – Plate 321	335
Grootblaarmirt – Plate 178	183
Grootblaarpieringbessie – Plate 380	395
Grootblaarrotsvy – Plate 180	185
Grootgeelbos – Plate 375	390
Grootkoorsbessie – Plate 324	338
Grootsuurpruim – Plate 159/196	163, 201
Growweblaarpieringbessie – Plate 142	145
Grysappel – Plate 218	225
Haak-en-steek – Plate 280	292
Hardekool – Plate 302	316
Hardepeer – Plate 95	97
Harige vingerblaar – Plate 353	368
Helikopterboom – Plate 288	302
Henkel se geelhout – Plate 135	138
Heuningnaboom – Plate 311	325
Hophout – Plate 7	8
Horingdoring – Plate 277	289
Horingpeultjieboom – Plate 335	350
Hottentotskersie – Plate 84	87
Huilboerboon – Plate 251	261
Huilboom – Plate 236	246
Jakkalskoffie – Plate 5	6
Jakkalsbessie – Plate 134	137
Kaapse boekenhout – Plate 51	54
Kaapse katjiepiering – Plate 231	240
Kaapse kiaat – Plate 146	149
Kaapse koffie – Plate 58	61
Kaapse kweper – Plate 133	136
Kaapse rooihout – Plate 359	374
Kaapse swarthout – Plate 66	69
Kaapse wilger – Plate 372	387
Kamassie – Plate 327	341
Kameeldoring – Plate 278	290
Kameelspoor – Plate 279	291
Karee – Plate 11	12
Karooboerboon – Plate 252	262
Kasuur – Plate 306	320
Keiappel – Plate 186	191
Kershout – Plate 363	378
Keurboom – Plate 242	252
Kiaat – Plate 291	305
Kieriekklapper – Plate 301	315
Kinaboom – Plate 138	141
Klapperbos – Plate 345	350
Kleinysterhout – Plate 163	166

Most Common Afrikaans Names

Kleingroendoring – Plate 194	199	Moerasvy – Plate 137	140
Kleinperdepram – Plate 14	15	Mopanie – Plate 285	297
Klipels – Plate 318	332	Muishondbos – Plate 8	9
Kliphout – Plate 187	192	Naaldblaarwitbos – Plate 326	340
Klipkershout – Plate 67	70	Namakwaharpuisboom – Plate 120	122
Klokkiesboontjieboom – Plate 284	296	Namakwavy – Plate 96	98
Klokkieskatjiepiering – Plate 188	193	Nanabessie – Plate 29	32
Knoppiesboontjie – Plate 272	284	Natalbosmelkhout – Plate 215	222
Knoppiesdoring – Plate 259	269	Natalappelkoos – Plate 167	170
Knuppelhout – Plate 336	351	Natalghwarrie – Plate 42	45
Koeboebessie – Plate 109	111	Natalkatjiepiering late 226	233
Koedoebessie – Plate 173	178	Natalkersielemoen – Plate 61	64
Koffiepeer – Plate 155	158	Natalklimop – Plate 300	314
Kokerboom – Plate 351	366	Natallukwart – Plate 197	202
Kokoboom – Plate 101	103	Natalmelkpruim – Plate 170	173
Koolhout – Plate 316	330	Natalmirt – Plate 74	79
Koorsboom – Plate 268	280	Natalokkerneut – Plate 189	194
Korentenaarpuisboom – Plate 65	68	Natalroosappel – Plate 76	80
Kosipalm – Plate 377	392	Natalse baakhout – Plate 344	359
Kraalnaboom – Plate 124	126	Natalse buksboom – Plate 171	174
Kransbessie – Plate 25	28	Natalse witstinkhout – Plate 80	83
Krantzvingerblaar – Plate 77	81	Natalvy – Plate 98	100
Kremetart – Plate 235	244	Natalysterpruim – Plate 185	190
Krinkhout – Plate 286	300	Nieshout – Plate 349	364
Kurkbos – Plate 241	251	Njalaboom – Plate 195	200
Kusrooimelkhout – Plate 157	160	Notsung – Plate 85	88
Kusswartbas – Plate 357	372	Olienhout – Plate 64	67
Kuswitessenhout – Plate 367	382	Omsambeet – Plate 254	264
Kwar – Plate 309	323	Omtisa – Plate 243	253
Laeveldkastaiing – Plate 368	383	Opregte geelhout – Plate 361	376
Laeveldse geelmelkhout – Plate 201	206	Outenikwageelhout – Plate 177	182
Laeveldvaalbos – Plate 373	388	Paddaboom – Plate 338	353
Lalapalm – Plate 227	234	Pambatieboom – Plate 56	59
Laventelboom – Plate 19	20	Papierbasdoring – Plate 266	276
Laventelkoorsbessie – Plate 151	154	Papierbaskanniedood – Plate 99/147	101, 150
Lebombo-ysterhout – Plate 312	326	Peperbasboom – Plate 210	217
Lebombokransesseboom – Plate 289	303	Perdepram – Plate 57	60
Lebombotrosblaar – Plate 293	307	Peulmahonie – Plate 267	277
Lekkerbreek – Plate 360	375	Platkroon – Plate 256	266
Lekkeruikpeul – Plate 271	283	Poeierkwasboom – Plate 221	228
Lemoenbos – Plate 50	53	Poerabessie – Plate 176	181
Lemoendoring – Plate 106	108	Pokysterhout – Plate 154	157
Lemoenhout – Plate 166	169	Pondoroosappel – Plate 75	80
Lepelhout – Plate 81	84	Pruimbas – Plate 114	116
Louriervy – Plate 97	99	Raasblaar – Plate 303	317
Maroela – Plate 211	218	Renosterkoffie – Plate 45	48
Meerstamvalsdoring – Plate 257	267	Reuserosyntjie – Plate 322	336
Melkpeer – Plate 216	223	Rivierverfbos – Plate 325	339
Mingerhout – Plate 365	380	Rondeblaargifboom – Plate 139	142
Mirtekweper – Plate 78	82	Rooi-ivoor – Plate 59	62
Mitseerie – Plate 111	113	Rooibitterbessie – Plate 140	143
Moepel – Plate 182	187	Rooiblaarrotsvy – Plate 83	86

411

Name	Page
– Plate 119	121
Plate 275	287
ate 364	379
nout – Plate 225	232
shout – Plate 363	378
weper – Plate 79	82
peer – Plate 33/60	36/63
ooistinkhout – Plate 30	33
Rotsessenhout – Plate 36	39
Saalpeultjieboom – Plate 331	345
Sambokpeul – Plate 330	344
Sandkroonbessie – Plate 68	71
Sandolien – Plate 295	309
Savanneboswilg – Plate 298	312
Sebrahout – Plate 238	248
Seeghwarrie – Plate 46	49
Sekelbos – Plate 281	293
Septeeboom – Plate 88	91
Shakamapruim – Plate 328	342
Silwerblaarmelkpruim – Plate 170	173
Silwerboom – Plate 376	391
Skurweblaarrosyntjie – Plate 108	110
Skurweblaarkanniedood – Plate 162	165
Smalblaarmosterdboom – Plate 55	58
Snotappel – Plate 193	198
Snuifkalbassie – Plate 228	236
Soetdoring – Plate 276	288
Springsaadboom – Plate 307	321
Stamvrug – Plate 164	167
Stekelblaarklapper – Plate 232	241
Sterkbos – Plate 292	306
Stinkebbehout – Plate 320	334
Stinkhout – Plate 379	394
Suurbessie – Plate 167	170
Swartapiesdoring – Plate 262	272
Swartbastertaaibos – Plate 37	40
Swarthaak – Plate 245	255
Swartklapper – Plate 229	237
Swartwortelboom – Plate 329	343
Swazi-oordeelboom – Plate 263	273
Sybas – Plate 69/315	72, 329
Tambotie – Plate 304	318
Terblans – Plate 371	386
Teerhout – Plate 356	371
Tontelhout – Plate 44	47
Tosselbessie – Plate 22	23
Trassiedoring – Plate 250	260
Treurboekenhout – Plate 370	385
Treurtrassiedoring – Plate 260	270
Tweelingbessieboom – Plate 165	168
Vaalboom – Plate 294	308
Vaalkameeldoring – Plate 283	295
Vaalrosyntjie – Plate 305	319
Vaderlandswilg – Plate 299	313
Valsdoring – Plate 264	274
Vanwykshout – Plate 237	247
Veldvy – Plate 54	57
Vlamdoring – Plate 240	250
Waterbessie – Plate 102	104
Waterpeer – Plate 161	164
Wilde mango – Plate 213	220
Wildeamandel – Plate 206	211
Wildeappelkoos – Plate 169	172
Wildedadelpalm – Plate 118	120
Wildefrangipani – Plate 340	355
Wildegeelkeur – Plate 239	249
Wildegranaat – Plate 362	377
Wildegroenhaarboom – Plate 269	281
Wildejasmyn – Plate 204	209
Wildekamperfoelieboom – Plate 191	197
Wildekastaiing – Plate 369	384
Wildekatjiepiering – Plate 231	238
Wildemirt – Plate 131	134
Wildemispel – Plate 179	184
Wildemoerbei – Plate 23	24
Wildenartjie – Plate 149	152
Wildeperske – Plate 192	197
Wildepietersieliebos – Plate 70	73
Wildepopulier – Plate 15	16
Wildepruim – Plate 160	163
Wildesering – Plate 247	257
Wildesuikerappel – Plate 219	226
Wildetulpboom – Plate 128	131
Witels – Plate 348	363
Witgat – Plate 82	85
Without – Plate 4	5
Witkaree – Plate 6	7
Witkatjiepiering – Plate 230	238
Witmelkhout – Plate 87	90
Witpeer – Plate 358	373
Witseebasboom – Plate 150	153
Witsering – Plate 343	358
Witstinkhout – Plate 26	29
Witysterhout – Plate 41	44
Wollerige kapperbos – Plate 208	215
Wolwegifboom – Plate 190	195
Worsboom – Plate 334	348
Ysterhout – Plate 163	166
Zoeloekersielemoen – Plate 122	124
Zoeloemelkbessie – Plate 117	119
Zoeloepeulbessie – Plate 168	171

Bibliography

Acknowledgement must be given to the authors, photographers and artists of the following publications as they proved invaluable references in helping me to accurately illustrate my workespecially those withered or partially dried specimens brought to me or after returning to my desk after a long tour away from home.

Adams, J., Wild Flowers of the Northern Cape (Cape Town: The Dept. of Nature & Environmental Conservation of the Cape of Good Hope, 1976)

Anderson, J., *Trees and Shrubs of the Witwatersrand, Magaliesberg & Pilansberg* (Cape Town: Struik Publishers (Pty) Ltd, 1988)

Author unknown, *The Bundu Book of Trees Flowers and Grasses* (Rhodesia: Longman Rhodesia, 1972)

Author unknown, *Trees and Shrubs of the Witwatersrand* (Johannesburg: The Tree Society of Southern Africa and Witwatersrand University Press, 1974)

Author unknown, *Wild Flowers of South Africa – National Botanic Gardens of South Africa, Kirstenbosch* (Cape Town: C.Struik Publishers, 1984)

Bohnen, P., *Flowering Plants of the Southern Cape* (Still Bay: The Still Bay Conservation Trust, 1986)

Codd, L.E.W., *Trees and Shrubs of the Kruger National Park* (Pretoria: Government Printer, 1951)

De Winter, B., M. De Winter and D.J.B Killick, *Sixty-Six Transvaal Trees* (Pretoria: Botanical Research Institute, 1966)

De Winter, B. and J. Vahrmeijer, *Die Nasionale Boomlys/The National Tree List* (Pretoria: J.L. Van Schaik Bpk, 1972)

Fox, F.W and M.E. Norwood Young, *Food from the Veld* (Johannesburg: Delta Books, 1982)

Funston, M., P. Borchert and B. van Wyk, *Bushveld Trees* (Cape Town: Fernwood Press, 1993)

Gibson, J.M., *Wild Flowers of Natal (Coastal Region)* (Pietermaritzburg: The Trustees of Natal Publishing Trust Fund, 1975)

Gledhill, E., *Veldblomme van Oos-Kaapland* (Kaapstad: Creda Press, 1981)

Hennessy, E.F., *South African Erythrinas* (Durban: The Natal Branch of the Wildlife Protection and Conservation Society of South Africa, 1976)

...E, C.L. Wicht and D.P. Ackerman, *Our Green Heritage. The South African Book* ...ape Town: Tafelberg, 1973)

Iwisisa, *Sappi tree spotting – Lowveld* (Johannesburg: Jacana Education, 1997)

, D. and S. Johnson, *Gardening with Indigenous Trees and Shrubs* (Midrand: Southern ...ok Publishers (Pty) Ltd, 1993)

...tham Kidd, M., *Cape Peninsula – South African Wild Flower Guide 3* (Cape Town: The Botanical Society of South Africa, 1883)

Moriarty, A., *Outeniqua Tsitsikamma & Eastern Little Karoo: SA Wild Flower Guide* (Cape Town: Botanical Society of South Africa, 1982)

Moll, E. and G. Moll, *Common Trees* (Cape Town: C. Struik Publishers, 1992)

Onderstall, J., *Sappi Wild Flower Guide, Mpumalanga and Northern Province* (Nelspruit: Dynamic, 1996)

Onderstall, J., *Transvaalse Laeveld en Platorand* (Cape Town: The Botanical Society of South Africa, 1984)

Palmer, E., *A Field Guide to the Trees of Southern Africa* (London: Collins, 1977)

Palgrave, K., *Trees of Central Africa* (National Publications Trust Rhodesia and Nyasaland, 1956)

Palgrave, K.C., *Everyone's Guide to Trees of South Africa* (Cape Town: C. Struik Publishers, 1975)

Palgrave, K., *Trees of Southern Africa* (Cape Town: C. Struik Publishers, 1977)

Sheat, W.G., *The A to Z of Gardening in South Africa* (Cape Town: C. Struik Publishers, 1982)

Timberlake, J. *Handbook of Botswana Acacias* (Gabarone: Ministry of Agriculture, 1980)

Thomas, V. and R. Grant, *Sappi Tree Spotting – Lowveld* (Johannesburg: Jacana Media, 1997)

Van Wyk, B, and P. van Wyk, *Field Guide to Trees of Southern Africa* (Cape Town: Struik Publishers, 1977)

Van Wyk, B.E. and N. Gericke, *People's Plants* (Pretoria: Briza Publications, 19897)

Van Wyk, P. *Trees of the Kruger National Park*, volumes 1 and 2 (Cape Town, Johannesburg, London: Purnell, 1972)

Von Breitenbach, F., Southern Cape Forests and Trees. (Pretoria: The Government Printer, 1974)